INJECTION MOULDING OF PLASTIC COMPONENTS

A Guide to Efficiency, Fault Diagnosis and Cure

INJECTION MOULDING OF PLASTIC COMPONENTS

A Guide to Efficiency, Fault Diagnosis and Cure

John Bown

McGRAW-HILL Book Company (UK) Limited

London · New York · St Louis · San Francisco · Auckland · Beirut
Bogotá · Düsseldorf · Johannesburg · Lisbon · Lucerne · Madrid
Mexico · Montreal · New Delhi · Panama · Paris · San Juan
São Paulo · Singapore · Sydney · Tokyo · Toronto

Published by
McGRAW-HILL Book Company (UK) Limited
MAIDENHEAD · BERKSHIRE · ENGLAND

British Library Cataloguing in Publication Data

Bown, John
Injection moulding of plastic components.
1. Injection molding of plastics
I. Title
668.4′12 TP1150 78–41107

ISBN 0–07–084521–2

12345 W.C.&S. 81079

PRINTED AND BOUND IN GREAT BRITAIN

This book is dedicated to my dear wife, Hazel, who cajoled, bullied, and encouraged me at all times into completing the script and on whose support I rely so much.

Contents

	Preface	x
1	**Introduction**	1
	Glossary and definition of terms	2
	Faults in injection moulded components	9
	Incidence and sources of faults	11
	Significance of faults	13
	Dealing with faults	16
2	**Principles of the process**	19
	The moulding cycle	20
	Plunger machine layout	23
	Plunger machine operation	23
	Inherent weaknesses	24
	Reciprocating screw machine	29
	Latest trends and developments in machines	32
	Understanding the process	33
	Developments in control systems	35
3	**Machines**	37
	The power unit	37
	Hydraulic fluid	40
	Hydraulic oil control	42
	Hydraulic pipework	43
	Machine sequencing	43
	Plasticizing unit	45
	Mould unit	52
	Machine specifications	59
	Machine control	61
	Process sequencing	63
4	**Materials**	65
	Thermosetting and thermoplastic materials	65
	Molecular or fine structure of plastics	66
	Chemical structure of plastics	68
	Polymerization and copolymerization	70
	Crystalline and amorphous plastics	71

Thermal properties of plastics 75
Effect of pressure on molten plastics 81
Mouldability of molten plastics 82
Rheological properties of plastics 83

5 Moulds 88
Mould design 89
Mould-making methods 95
Gates for injection moulds 96
Calculation of gate sizes 98
Runner systems for injection moulding 100
'Family' moulds 102
Runner dimensions 102
Three-plate and hot runner moulds 102
Extended nozzles 104
Mould cooling systems 105
Differential shrinkage and cooling effects 105
Component design 106

6 The process and its control 108
Determination of injection rate 109
Determination of power requirement 112
Pressure requirements for injection 112
Hydraulic pump 115
Trimming the process 116
Balancing process variables 117
Switching systems and process variables 119
Mechanical systems and process variables 119
Feed setting 120
Interdependence of process variables 121

7 Injection moulding of thermosetting materials 122
Problems with thermosetting materials 122
Machines for thermosetting plastics 123
Thermosetting moulding materials 124
Process details 125
Mould design for thermosetting moulding 127
Faults in thermosetting mouldings 128

8 Causes, diagnosis and cure of faults in injection mouldings 130
Burn marks 130
Cracking or crazing 132
Delamination 132
Distortion 133
Flash 133

	Flow marks	134
	Poor surface finish	135
	Scuffing	136
	Short mouldings	136
	Sink marks and voids	138
	Splash marks, mica marks, silvering	139
	Splay marking	141
	Staining	141
	Sticking	142
	Warping	143
	Weld lines	144
	Windows	146
	Witness marks	146
	Other faults and problems	146
9	**Housekeeping as a factor in efficiency**	149
	Cleanliness	149
	Contamination and waste	151
	Safety aspects of housekeeping	153
	Maintenance	156
	Plant interior decoration and colour coding	157
10	**Keeping records**	159
	Why keep records?	159
	Material balances	163
	Machine operation records	165
	Job records	166
	Mould records	166
11	**Factory layout, services and ancillary equipment**	168
	Power requirements	168
	Power factor correction	169
	Disposition of machines	169
	Service lines	171
	Cooling water supply	172
	Mould temperature controllers and chillers	176
	Granulating and mixing equipment	177
	Hopper loaders	178
	Hopper driers	178
	General drying equipment	179
	Raw material and finished product stores	180
	Treatment and care of re-work	181
12	**Financial aspects**	182
	Fixed and variable costs	183

Budgeted machine hourly rates 188
Estimating and pricing of component production 189
Cost saving methods 190
Optimizing time cycle and machine productivity time 193
Effects of process variables on profitability 197
Conclusion 200

Index 203

Preface

Much has been written about the injection moulding of plastics, both in books and in articles. Nearly all manufacturers of raw materials have produced brochures on the subject, explaining the peculiarities of their products, and there are treatises ranging from a simple learning script to highly complicated mathematical analyses. In this context, why should another book be necessary?

There are, in the United Kingdom, perhaps one hundred companies engaged in large scale injection moulding, but there are also many hundreds of firms who operate on a very small scale, some with one or two machines only. It has been my privilege, over the last ten years or more, to meet many people who are interested in the subject. Some have been moulders, some mould makers; others have been machinery manufacturers and there have been many whose interest has been solely as users of moulded components. All have agreed that the injection moulding of plastics is an intriguing subject of absorbing interest. The degree of complexity has been likened to the moves of a knight in three-dimensional chess!

To these people I offer my sincere thanks for all the help they have given me in increasing my understanding of the injection moulding of plastics. This book is an attempt to place on record some of the things I have learned in twenty-five years of study of the subject. It is not intended as a guide to enable an engineer to build a machine nor for a fitter to mend one. The intention has been to make an analytical study to try and understand why things go wrong.

It has been wisely said that accidents do not happen: they are caused. Sometimes by ignorance, occasionally by stupidity. It is the same with injection moulding. Troubles and faults are not accidental; they occur because someone or something has caused them. In this book, some of the causes are examined and discussed with a view to banishing some of the ignorance. Stupidity cannot so easily be catered for!

I have found in the past that some of the writings on injection moulding have been very erudite and full of mathematical discussions but they have often failed to make concrete suggestions that could be understood at shop-floor level (not that shop-floor level is, in these days, by any means 'low'). In this book there is some mathematics, simple, let it be said, and reduced to terms that all should understand. Mathematical purists may scorn this simple approach, but I have found that presenting the facts in this way has enabled those who, like me, are not mathematical geniuses, to have a new understanding of the subject. It enables some definite, concrete proposals to be made. That these proposals work in practice can hardly be disputed and, in fact, where they have been tried out it has been possible to improve performance in astonishing ways.

In an industry where there is increasing need for efficiency, any means which will help to remove causes of waste must be welcomed. Some measures such as spending money on cleaning or putting extra effort into keeping records are probably unpalatable and other ideas may never have been attempted because of the consequences of failure. This book is offered with these things in mind. If it also helps those who are just beginning their study of the technology of injection moulding that will, indeed, be a bonus.

I express my grateful thanks to all who, knowingly or not, have assisted in the preparation of this book. Some, whose work I have quoted or called on, I have credited in the text. To my good friends, John R. White and John E. Nightingale, I give special thanks for all their help.

John Bown
North Runcton
Norfolk

Acknowledgements

The Author offers grateful thanks to the following:

Imperial Chemical Industries Limited, Plastics Division, for permission to use Figs. 1.1, 1.2, 1.3, 1.4, 2.2, 2.3, 2.4, 2.5, 3.3, 3.4, 4.8, 4.9, 4.10, 8.1, 8.2, 8.3, 8.4 and 8.5.

GKN Windsor Limited, for permission to use Fig. 2.8.

Sandretto Plastics Machinery Limited, for permission to use Fig. 2.9.

1

Introduction

Wherever one cares to look around in the home, the office, the factory, in industry, agriculture, horticulture or, in fact, in all aspects of modern life, one sees plastic injection mouldings. Their applications are numberless and, for the most part, their performance is unexampled. But they all have to be produced from granular or powdered plastic by a quite complicated process, and although the process is designed for mass production, all along the line there are difficulties. Each moulding should be exactly like every other moulding produced from a particular mould, but in order to ensure that this is the case many problems have to be overcome. This book is an attempt to approach the subject of the injection moulding of plastics in an analytical manner, to consider the problems that are encountered, and to discover their causes. Having done this, the method by which they may be overcome will be worked out from the basic principles of the process.

A close study of any common household article which is made from plastic by injection moulding—say a jug or basin, an electrical plug and socket, a radio cabinet or a vacuum cleaner case—will reveal many areas where problems can occur. Glossy surfaces must be unblemished, white items must not be contaminated with black specks (nor black articles with white specks), and handles must be smooth and edges not sharp enough to cause injury to the user. The design and production of the article must be such that no untimely breakdown occurs and it must be suitable for its purpose. Such matters present a challenge to the practical moulder, to whom a large section of this book is addressed.

In meeting this challenge it has to be remembered that a business cannot function unless it is financially viable and produces a useful return on the amount of capital employed. Because the process is 'capital intensive', requiring more capital per person employed than do many industries, the financial aspects of injection moulding have not been forgotten. With the increasing interest by the 'shop floor' workers in the proper running of a company, it does not seem out of place in a book which is, after all, aimed primarily at the technical side of the business, to devote a chapter to the financial aspects of injection moulding.

Health and safety at work are assuming an increasingly important role, and this has not been neglected in the book. Although the chapter which deals with this is primarily concerned with housekeeping and efficiency, the relationship between health, safety and efficiency cannot be too strongly emphasized. Unfortunately, it has to be admitted that the injection moulding industry is not among the best so far as housekeeping and cleanliness are concerned and it may be opportune to take a self-analytical look at this subject and to see what the effects of poor housekeeping are likely to be.

Plastic injection moulding is a process which depends for its success on the proper functioning of a *machine* which is operating with *material* and producing the shapes required by means of a *mould*. Each of these three elements is important and the interdependence of machine, material and mould will become more apparent as the work proceeds. In order thoroughly to understand the functions of each it is necessary to know the jargon of the business. Just as every industry from pottery to ironfounding has its own language, so injection moulding has words, phrases and terminology that would not be readily understood, or may have a different meaning, outside the context of the injection moulding shop. Studying some of these terms will therefore be the first consideration, making sure that, so far as is possible, terms are defined in as concise and unambiguous a manner as possible. Reference should be made to British Standard 1755: Glossary of Terms Used in the Plastics Industry for any further information, but the terms used and defined in this book are those which, by experience, have been found to be in most common use among many people, at all levels of technology, who work in the industry. Many glossaries are placed in an appendix at the end of a book. Because of the importance of knowing the language this glossary is included in the introductory chapter and is in two sections: (1) Glossary and definitions of terms; and (2) faults in injection moulded components.

Glossary and definition of terms

Accumulator	A device for storing hydraulic fluid under pressure.
Appearance surface	The surface of a moulding which will be seen and which therefore has to present a suitable appearance.
Back-flow stop valve	A non-return valve on the tip of the screw which prevents fluid plastic from passing backwards along the screw flights when injecting.
Backing plate	A supporting plate in the mould carrying the cavity blocks (*see also* Bolster).
Balance tray	A tray under the mould on to which the moulding falls during automatic cycling and which then operates the switch to close the mould and so causes

	the re-cycling starter to operate to re-start the cycle. Used instead of a pause timer and as a mould-safety device.
Barrel	*See* Cylinder.
Bolster	A block of metal (usually steel) for carrying cavity blocks in an injection mould. Also called a 'chase'.
Carrot	*See* Sprue.
Cartridge heater	An electrical resistance heater in the form of a cartridge for inserting into a nozzle or a section of a mould.
Cavity	The hollow space in the mould where the plastic is formed to shape.
Cavity register	Angled faces on parts of the mould which match when the mould is closed and so ensure correct alignment of the parts.
Check ring	A ring type of non-return valve on the tip of the screw which prevents fluid plastic from passing backwards along the screw flights when injecting.
Compound	A mixture of plastic polymer and additives which has been fused by heating and then granulated or otherwise prepared for feeding to a moulding machine.
Compression ratio	The ratio of the volume of one flight of a screw (i.e., one complete revolution) at the feed end to that at the discharge end.
Copolymer	A polymer made by polymerizing together two or more different monomers.
Core and core pin	The solid part of the mould which gives the inside shape to the moulding.
Cuff heater	*See* Heater band.
Cylinder	The cylindrical part of the plasticizing unit of an injection moulding machine.
Cycle time	*See* Moulding cycle.
Daylight	The working distance between the fixed and moving platens of an injection moulding machine.
Decompression zone	In a vented barrel, the decompression zone comes in between the first and second compression zones and allows venting of volatiles without escape of plastic (*see* Vented barrel).
Degating	Separating the mouldings from the runner system.
Depolymerization	*See* Polymerization.
Dice cut	Material for processing in the shape of cubes of approximately 2–3 mm in size.
Dowel pins	Guide pins of hardened steel which hold one part of a mould in the correct position relative to the other

	part or parts. The guide pins fit closely into bushes made of hardened steel.
Draft	The taper on parts of a mould designed to allow easy extraction of the moulding from the mould.
Draft angle	The angle of the taper from the vertical (i.e., from the direction of mould opening).
Ejector pin	A steel rod which moves forward as the mould opens, pushing the moulding from the mould.
Ejector plate	A plate in the mould which carries the ejector pins.
Ejector system	A device fitted to the moving platen of the machine for operating the moulding ejectors. It may be operated mechanically (including springs), hydraulically, pneumatically or electrically.
Extended nozzle	An extension of the cylinder nozzle which penetrates into the mould and shortens, or obviates the need for, a sprue bush.
Flash mould	A *horizontal* flash mould has the mating surfaces at right-angles to the direction of opening of the mould. A *vertical* flash mould or positive mould has the mating surfaces in line with the direction of opening (*see* Fig. 5.2).
Flow moulding	A method of injection moulding in which the screw continues to rotate during injection, thus enabling larger shot weights to be obtained than would normally result. The injection rate is much slower than in normal operation and the process is generally applicable to the moulding of thick sections.
Gate	A restriction between the channel conveying material to the cavity and the cavity itself. For types of gate *see* Chapter 5.
Heater band	An electrical resistance heater in the form of a cuff which encircles the cylinder or nozzle.
Impression	A mould cavity. Moulds may be designated single-impression or 2-, 3-, 4- etc. or multi-impression.
Injection moulding	The process whereby a quantity of plastic material in solid or granular form is heated and softened in one part of the machine and then forced under pressure into a mould where it is cooled and hardened (or heated and cured in the case of thermosetting materials) and retains the shape imparted to it.
Injection pressure	The pressure exerted on the material in the cylinder by the ram or screw during injection. Injection

pressure = line pressure × intensification ratio (q.v.).

Insert	A piece of material such as a screw or a stud placed in the mould before injection takes place and around part of which the plastic is moulded so that the insert becomes an integral part of the component.
Intensification ratio	The ratio of the injection pressure to the pressure of the hydraulic fluid (line pressure). It is numerically equal to the area of the hydraulic cylinder which actuates the screw or plunger divided by the area of the plunger itself.
Jig cooling	Cooling a moulding, after removal from the mould, in a jig which prevents it from changing shape.
Land	Narrow rims in a mould where parts of the mould 'mate' together. Also, the parallel parts of a gate.
Lift	A moulding, or set of mouldings, produced in one cycle of the machine.
Limit switch	A switch set at the limit of movement of part of the machine which controls the electrical power supply either to stop further movement of that part or to initiate a further part of the moulding cycle.
Line pressure	The hydraulic pressure in the supply pipes (or lines) to the part of the machine being used.
Locking force	The force exerted in the locking mechanism of the machine which keeps the mould closed during injection.
Locking mechanism	A hydraulic cylinder or toggle mechanism which is designed to close the mould and to keep it in the closed position during injection.
Masterbatch	A concentrated mixture of pigment, filler, antistatic agent, lubricant or other additive, and a base polymer (e.g., polyethylene, polypropylene, etc.) This is added by the moulder, as required, to the moulding material to give the concentration necessary to produce the colour or other quality desired.
Monomer	A chemical compound consisting of simple molecules which can be made to join together by a process called polymerization to produce a polymer which is composed of much more complex molecules and is the basic material from which plastics are made.
Mould	A hollow form or cavity into which molten plastic material is forced to give the shape of the required

component. The term generally refers to the whole assembly of parts which go to make up the section of the injection moulding equipment in which the plastic components are formed. Also referred to as the 'tool'. For various types of mould, *see* Chapter 5.

Mould height	The overall thickness of the mould as it is located between the platens of the machine.
Moulding cycle	The complete sequence of operations necessary to produce a moulding or set of mouldings. Also, the total time taken to complete the sequence (cycle time).
Non-return valve	A one-way valve in the hydraulic system or on the tip of the screw (*see* Check ring).
Nozzle	A constriction on the front of the cylinder, which has the effect of reducing the diameter of the cylinder to that of the sprue bush, with which it can then form an effective seal, and through which material passes into the mould channels (*see also* Extended nozzle). For types of nozzle, *see* Chapter 5.
Pause timer	The process timer which starts the closing of the mould ready for the next cycle to begin. Used when the machine is operating fully automatically.
Plasticize	To soften plastic material either by the incorporation of a high boiling point liquid (a plasticizer) or by heating the material with or without mechanical mixing.
Plasticizing capacity	The amount of material which can be plasticized ready for moulding by heating in the cylinder of an injection moulding machine in a given time. Expressed as lb/h or kg/h.
Plasticizing cylinder or unit	The part of the injection moulding machine where the material is softened (or plasticized) prior to being injected into the mould.
Platens	The steel plates on the mould end of the machine to which the parts of the mould are attached. The 'fixed' platen is that nearer to the cylinder and the 'moving' platen, as implied, can move to open or close the mould.
Plunger	In small or old-type machines, the plunger is a steel rod which is a clearance fit to the cylinder in which it reciprocates to bring about injection.
Polymer	A chemical compound made up of a large number of repeating structural units (*see also* Monomer).
Polymerization	The process whereby a monomer (q.v.) is turned

	into a polymer (q.v.). The reverse process, called depolymerization, may occur when a plastic is heated, liberating volatile monomer.
Preheating	The process of heating material prior to feeding it to the cylinder. This can take place either in the feed hopper or in an external heater.
Process timer	A clockwork or electronic timing device for switching an electrical circuit at a predetermined time.
Projected area	The area (of a moulding) as projected on to a plane at right-angles to direction of opening of the mould in which it was made.
Ram	*See* Plunger.
Register ring	A ring encircling the sprue bush on the mould which is a close fit into the hole in the fixed platen of the machine and which 'registers' the centre of the sprue bush to the centre of the nozzle. It assists in accurate setting of a mould on the platen.
Relay	A mechanical/electrical device in which a small electrical current is made to switch a large one.
Runner	The channel in a mould which connects the sprue bush to the gate.
Screw	The screw of an injection moulding machine consists of a round bar of hard steel in which is cut a helical channel (screw thread). It rotates inside the cylinder conveying material from the feed pocket to the nozzle and can also move forwards or backwards to effect injection or to allow material to accumulate at the nozzle end of the cylinder (*see* Chapter 3).
Setting	The preparation of a machine ready for moulding. It involves fixing the mould in place and making all the adjustments necessary to ensure that the machine works properly and efficiently.
Setting up	A term sometimes used to denote the hardening of material in the mould prior to removal of the moulding.
Shot capacity	The amount of material which can be injected during one cycle of the machine (*see also* Swept volume).
Shrinkage	Also called 'after-moulding shrinkage', this is the reduction in dimensions which a component undergoes after moulding. It is expressed as a percentage or as inches per inch (mm/cm). The component is assumed to start with the same dimensions as the mould.

Shut-off nozzle	A nozzle which incorporates some kind of valve which prevents leakage from the nozzle.
Solenoid	An electromechanical device in which an electric current causes an electromagnet to move. This can be used in injection moulding for opening or closing a hydraulic valve.
Sprue	The solid material which forms in the sprue bush (q.v.), sometimes referred to as the 'carrot'.
Sprue bush	The inlet channel for material into the mould or runner system cut through the fixed plate of the mould.
Sprue break	After injection and screw-back, the nozzle may be moved back from the sprue bush to give a small gap during the period when the mould is open. This process is called 'sprue break'.
Stripper plate	A plate that moves over part of the male portion of the mould and which slides forward as the mould opens, pushing the moulding off the core.
Swept volume	The volume of material which is displaced as the screw or plunger moves forward. It is the effective area of the screw or plunger multiplied by the distance of travel.
Terpolymer	A copolymer, such as ABS, made from three monomers. ABS is made from acrylonitrile, butadiene and styrene.
Thermocouple	A device for measuring temperature, consisting of two pieces of electrical resistance wire of different types welded together at one end. The wires are insulated from each other and connected to a sensitive millivoltmeter. A small current is generated in the system and this is proportional to the temperature at the weld.
Tie bar	A stout steel bar, one of two, three or four, connecting the fixed platen of the machine to the locking mechanism and on which the moving platen is carried.
Tolerance	The permissible latitude in a specified quantity or dimension.
Torpedo	In a plunger machine, the solid metal block of streamlined form fitted inside the cylinder which causes the molten material to flow near to the cylinder walls. Also called a 'spreader'.
Vent and venting	Means by which air can escape from the mould as it is filled.
Vented barrel	A plasticizing cylinder which has a vent about half-

way along its length, through which volatile matter can escape. Used in conjunction with a special screw having two separate threads along its length connected by a zone where decompression of material takes place.

Faults in injection moulded components

(Figures 1.1 to 1.4 show examples of common faults)

Burn mark — An area of degraded or oxidized plastic in a moulding.

Cracking or crazing — Fine cracks on the surface of a moulding which sometimes turn into deeper cracks when stress is applied.

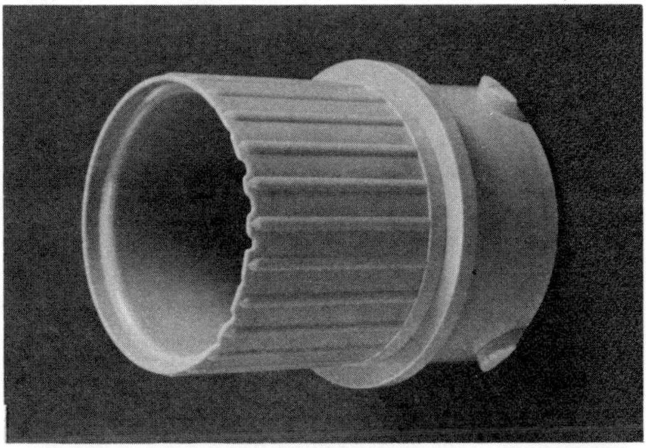

Fig. 1.1 Short moulding

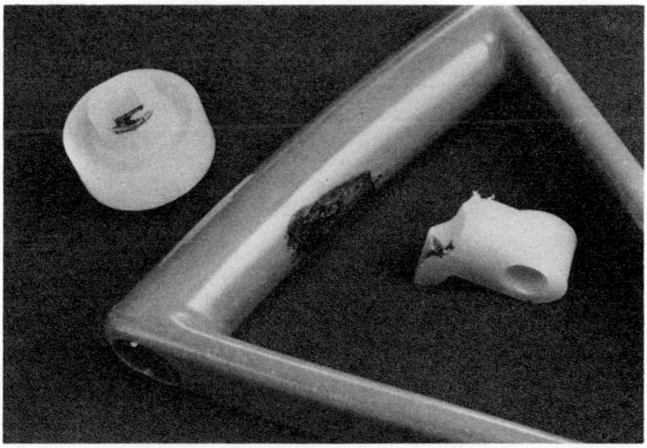

Fig. 1.2 Badly burned mouldings

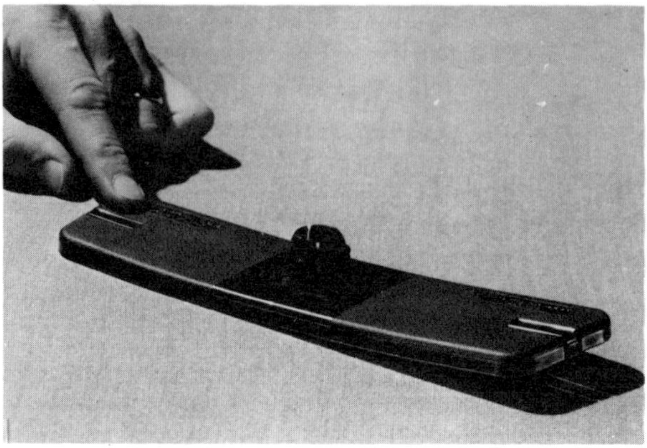

Fig. 1.3 Warped moulding

Delamination — Surface layers of material which separate from the main section of a moulding; in appearance, rather like the skin of an onion.

Distortion — The moulding is of a different shape from that required, generally apparent immediately the component leaves the mould (*see also* Warping).

Flash — An unwanted thin web of material on the edges of a moulding or at mould mating surfaces where gaps should normally be too small for material to penetrate.

Flow mark — A mark on, or in, the surface of a moulding which shows the flow path of the material, especially where the stream of material has become divided (*see also* Weld line).

Mica mark — *See* Splash mark.

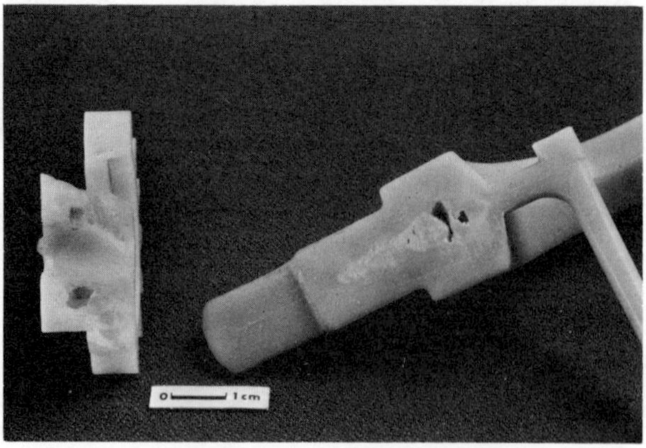

Fig. 1.4 Void in thick section of moulding

Poor surface finish	Undesired dullness or ripple on the surface of a moulding.
Scuffing	Surface marks on the walls of a moulding which run in the direction of mould opening.
Short moulding	An incomplete moulding.
Silvering, silver streak	*See* Splash mark
Sink mark	An unwanted depression in the surface of a moulding, especially on the face opposite to where the section thickness increases (e.g. at a rib).
Splash mark	A mark on a moulding, having an appearance similar to mica. Splash marks generally (but not invariably) radiate from the point of entry of material into the mould. Also known as mica mark, silvering or silver streaking.
Splay mark	Surface marking on the moulding, generally near to the gate, and similar in appearance to the grooves of a gramophone record.
Staining	Areas of a moulding (or mould) which take up a dull, discoloured appearance.
Sticking	The moulding tends to stick in the mould by mechanical or vacuum adhesion of the moulding to the mould surface, making ejection difficult. Sometimes this fault only occurs at the sprue.
Voids	Only obvious in transparent mouldings, voids are unwanted cavities within the thickness of the material. They occur also in opaque mouldings, when they can be detected either by ultrasonic means or by density determination (or by a destructive test).
Warping	The moulding becomes bowed or distorted, generally some time after cooling (*see also* Distortion).
Weld line	A mark on the surface of a moulding at the junction of separate streams of material.
Window	An area of transparent material in an otherwise translucent moulding.
Witness mark	A mark or line on a moulding where different parts of the mould come together or 'mate'.

Incidence and sources of faults

A careful scrutiny of almost any article injection moulded in plastic will reveal at least one minor defect. Some defects are so small as to be insignificant. Others may be quite easily seen and identified, even by an untrained person. All too often, a component may have a fault which makes it virtually, if not actually, unusable and a disfigurement of an 'appearance surface' is quite common. Unfortunately, far too many household articles have bad moulding faults and most housewives can tell of washing-up bowls that 'peel', of buckets that split across the bottom and of moulded boxes with ill-fitting or distorted lids. It is a strange observation that many 'throw-away' items are excellently moulded, while numerous articles intended for permanent use have quite bad faults. For example, it was found that the polyethylene cap for use on a tin of coffee, supplied free for sealing the contents after opening the tin, was an almost perfect moulding but a polythene box bought for use in the larder for food storage had a badly distorted lid and the sides of the box itself were badly bowed. Could it be that the moulder who made the good lid did not know the reason for his success and that the one who made the bad box did not know what was being done wrongly?

The importance of machine and material. Faults in injection mouldings may be due to a number of causes arising from the different elements of the process. Poor machine design or machine malfunction for one reason or another is a fruitful cause of faulty component production. This aspect is dealt with in Chapter 3. In Chapter 4 the materials for injection moulding are discussed, including some of the main thermosetting plastics now being used. Injection moulding is more particularly suited to the thermoplastic materials, typified by polyethylene and nylon, but the behaviour of even these materials is far from simple. Variations of pressure, temperature and of the rate of application of shear stresses are all inter-related and an alteration in one parameter will interact with all the others.

It has to be remembered, fundamentally, that plastics do not consist of small, simple, unconnected molecules, as do metals, but of complicated, long-chain molecules which can be drawn out, entangled or otherwise physically changed in a much more complex way than the molecules of metals.

The importance of the mould. The third element, the mould, is discussed in Chapter 5. Mould design, as a subject, is not easily taught or learned because of the almost infinite variation possible, and it is unfortunately true to say that most mould designers have had to learn by a process of trial and error. It is not the function of this book to deal with the subject comprehensively but the basic elements will be described, the reader then being referred to some of the excellent books available. However, it must be observed that the component designer may occasionally call for a mould to be made in which some of the really important principles of good mould design have to be disregarded, resulting in a mould which is most difficult to operate. Because by the time a mould has been designed and made a great deal of money may have been spent and no further funds are forthcoming, the moulder has the job of trying to make good mouldings in a poor mould. Some of the expedients which can be used in such a case will be mentioned but it cannot be too strongly or too frequently stated that liaison between all parties concerned at all stages of a project can often obviate these kinds of difficulties.

The all-important process. In Chapter 6 the three elements of injection moulding come together in the discussion of the process and of process control. An incorrect balance of the functions which make up the process—time, temperature and pressure—can result in faulty mouldings and loss of productive output.

Causes, diagnosis and cure of faults. Because of the differences in behaviour during injection moulding, Chapter 7 is devoted to thermosetting materials, reverting again to thermoplastics in Chapter 8, which looks into the causes, diagnosis and cure of faults in components and includes a check-list compiled as a quick reference to the parts of the book dealing with the various aspects concerned. Some case histories are detailed and an explanation is given of a simple system of diagnosis.

As mentioned earlier, housekeeping is a factor in efficiency as well as being an important item in health and safety at work. Many of the faults that occur in injection moulded components can be directly attributed to bad housekeeping practices and the importance of cleanliness in working conditions is emphasized in Chapter 9. Legal aspects of these matters are outside the scope of this book and the reader should consult the publications of the Safety Council and The British Plastics Federation.

The subjects of the following two chapters may not appear immediately to be relevant to faults and fault diagnosis. It is pointed out in Chapter 10, however, that the lack of good records may often make it impossible to ascertain the causes of faulty mouldings; whereas with adequate, though not necessarily over-elaborate records, these causes may more readily be discovered. Efficient factory services and ancillary equipment, discussed in Chapter 11, can also be a contributory factor in reducing the incidence of faults in mouldings. Finally, many of the financial aspects of injection moulding, discussed in Chapter 12, are very closely linked with the incidence of faults in the components being manufactured.

Faulty component production, then, is one of the most severe problems in the injection moulding of plastics. The production of too large a percentage of mouldings which fail to pass the inspectorate can be a severe limitation to profitability, and consideration needs to be given to the question, 'what constitutes a fault?'.

Significance of faults

Many of the faults that occur in injection moulded components may have little effect on the performance of the article or on its sales appeal. Nevertheless, a keen inspectorate will often reject them for some reason, claiming that they do not conform to specification. What, then are the criteria that should be laid down in specifications and that need, therefore, to be considered?

Specification criteria. The first criterion is, 'does the fault impair performance?' Some faults, such as poor surface finish or sink marks may not be detrimental to performance whilst a short moulding or a moulding having voids in a thick section may either be completely unusable or it may fail in service. For example, a photographic film cassette having the most minute pinhole caused by producing a short moulding would be entirely useless. On the other hand, a 'throw-away' moulding such as is used in packing office machines for despatch could be quite usable for its purpose if it showed large sink marks. A component having a void in a section which comes under stress is liable to fail in service because the void provides a region of discontinuity of the plastic and hence of increased stress, in other words, a 'notch'.

The second criterion is, 'does the fault cause a reduction in sales appeal and make the product unsaleable?' Many components are sold on their sales appeal and a defective 'appearance surface' makes the article a reject. Typical examples are clock or watch 'crystals', many household articles, and car components. If one of the main functions of the component requires a perfect surface, free from all blemishes, then this must be the main criterion for production and inspection. On the other hand, a blemish on the surface of an article which is not going to be seen or even such an imperfection as poor colour dispersion is not going to affect the sales of the complete assembly and therefore need not be a reason for rejection of a moulding.

Perhaps a third important criterion is the dimensional tolerance allowed in the component. When a mould has been proved and the moulded components have been passed by inspectorate it would be imagined that, provided conditions were kept steady the dimensions of the articles produced would be identical. In practice, this seldom occurs and, for reasons which will be discussed later, small deviations are found. However, the degree of accuracy which is expected may be unrealistic in the first place because dimensional tolerances may have been set without reference to one of the most important properties of plastics, i.e., the thermal expansion of the material. To an engineer used to working with metals with fairly small coefficients of thermal expansion, the allowances required for plastics may well seem unacceptable, especially when the average thermal expansion is expressed as, for example, 0·01 per cent/°C. This means that a component 10 in (25 cm) long over a range of $-20°C$ to $+60°C$ would change in dimensions by 0·08 in (2 mm), so that the setting of a moulding tolerance on a length of $\pm 0{\cdot}002$ in (0·05 mm) would seem a little unrealistic.

What, then, constitutes a 'fault' in a moulded component? In practice, a faulty moulding is one which will not pass the scrutiny of the inspector, but the criteria of the inspectorate are often laid down with little or no reference to the actual needs of the user, nor to the standard expected of the moulder. Conflict arises and rejected mouldings accumulate until the profitability of the particular job is reduced to an unacceptable level. It is advisable to make sure that realistic inspection criteria are agreed at the start of a project so that no one expects the impossible.

The importance of dimensional tolerances. It must be appreciated that extremely minor faults can be cumulative. Two examples of this may be cited:

In the manufacture of photographic colour slide mounts the overall thickness is rather less important than is the variation in thickness across the width of the mount. A 2 per cent variation in thickness, say $\pm 0{\cdot}000\,75$ in (0·02 mm), means that a stack of mounts as fed to the automatic mounting machine would vary by just under half an inch (10 mm) and would cause problems in mounting the transparencies.

In the manufacture of concrete-casting moulds used in the construction of large buildings, a small variation in the overall width of the moulds causes them to be rejected because, it is said, the finished building would be of incorrect size. (Perhaps it should be mentioned, however, that the dimensional tolerances demanded would be offset by a temperature change of 20–30°C).

Setting dimensional tolerances is one of the most difficult exercises for the designer. In addition to thermal expansion, the degree to which moisture is absorbed may also be of importance. An example of this occurred in the manufacture of drive bushings in nylon type 66 where the mouldings produced were brittle. (This is general with some grades of nylon because toughness is only imparted to the material by the plasticizing effect of the $2\frac{1}{2}$–4 per cent of water absorbed at equilibrium (at 65 per cent relative humidity) and, of course, no moisture can be present during moulding.) Conditioning in boiling water for 30 min reduced the brittleness but caused the bore to close slightly so that the spindle would not easily fit. The bore was then reamed to size, but when the bushing had absorbed its full equilibrium content of water, all the dimensions had grown so that the spindle became too easy a fit. Such cases are rare, but have been observed, and generally could be avoided by a little pre-knowledge and anticipation. Of course, the complete answer is to have a large enough stock of components so that they can be brought naturally to equilibrium moisture content over a period of several weeks.

In another instance there was a more fortunate outcome. Gears for counting equipment, designed and moulded in nylon, failed in service in humid, tropical conditions because the combined effect of heat and moisture caused them to become too large and to seize up. The moulder had a quantity of acetal copolymer, which does not absorb moisture, and supplied samples in this material with the result that the nylon could be replaced directly. This practice could not always be followed, however, unless the conditions of service would allow it and it is always better to decide on the appropriate material and design correctly to allow for expected dimensional changes. Because some design engineers are averse to innovation one example of failure of a component in service will often undermine confidence in the ability of 'plastics' generally to perform satisfactorily.

Surface finish. One other common cause of the rejection of mouldings by inspectors is poor surface finish. Although this can be of great importance, as has already been stressed, where the component is going to be seen, in a moulding which will be hidden, say behind the facia panel of an automobile or in the interior of a piece of domestic equipment, the surface appearance does not matter so much. In general, if properly moulded, plastics will take up the exact surface finish of the mould. Producing a high quality finish on a mould is expensive and should never be done unnecessarily.

Dealing with faults

Most moulders and other personnel connected with plastic moulding are familiar with fault-finding charts which include 'causes of fault' and 'possible cures or suggested remedies'. Such charts have a certain limited use, but in general they have much too diffuse or 'sloppy' an approach to be of great value. Part of a typical chart could be as follows:

FAULT	POSSIBLE CAUSE	SUGGESTED REMEDY
Sink marks	Material too hot	Reduce cylinder temperature.
	Insufficient material injected into mould	Increase feed. Increase cylinder temperature. Increase mould temperature. Adjust gate size.
	Insufficient injection time	Increase injection time.
	Gate freezing prematurely	Increase gate size.
	Gate located incorrectly	Change position of gate.
	Mouldings too hot when ejected	Increase cooling time; reduce mould temperature; drop mouldings into cold water.
	Effective pressure in cavity too low	Increase injection pressure. Increase cylinder temperature. Increase mould temperature. Alter position or size of gate.

. . . and so on . . .

Such charts are virtually useless. In the space of a few lines the chart indicates: (a) the cylinder temperature is too high; (b) the cylinder temperature is too low; (c) the mould temperature is too low; (d) the mould temperature is too high; and (e) the gate is in the wrong place or the mould needs to be radically altered. In other words, the poor moulder is really being told, 'try everything and see whether it works; sooner or later you may hit on the right answer.'

This sort of advice is really not good enough. The properties and behaviour of plastics are sufficiently well known to enable a much more scientific approach to be made to fault-finding. In Chapter 4 some of these properties are discussed more fully and later in the book are more closely related to the causes and cures of faults. Reverting again to the causes and cures of sink marks instanced above, the preferred approach is based on the changes in volume which occur in the material during heating and cooling.

The specific volume of plastics increases by between 0·05 and 0·11 per cent/°C rise in temperature. The former figure is typical of amorphous plastics, the latter of crystalline ones. Over the temperature range from moulding temperature to room temperature the molten material reduces in volume by between 10 and 20 per cent. Thus, if a mould is exactly full at moulding temperature, the material will 'sink' away from the walls of the mould as the material cools, the effect being greatest where the section is thickest. The result is 'sink marks'. Some means must therefore be found of increasing the actual *mass* of material filling the mould and this includes increasing the pressure, making the material flow more easily, or forcing more material into the mould as the volume of material in the mould decreases. If the basic reason for 'sinking' is known and the various parameters controlling mould filling are understood, the causes and curing of sink marks will become apparent. Let it be said here, however, that altering the mould temperature or the cooling time will not make any difference to the amount of material entering the mould and can only serve to change a sink mark into a void. The approach in Chapter 8 where a check-list of faults, causes and cures is presented involves, as outlined here, a thorough understanding of exactly what is happening.

Another very common piece of advice given in some fault-finding charts is 're-design the mould' or 'alter the position of the gate'. Unfortunately, to follow this advice may prove extremely time consuming and expensive. Any alteration to the mould, other than of a very minor nature, should be made only as a last resort. By the time a mould is ready to put on to a machine a great deal of money has already been spent. It has been designed by someone who has thought out all the aspects of its possible functioning and has been made by a mouldmaker who is, to some extent, skilled 'in the art' and if the work has been done correctly, it should be possible to make it work. In any case, the person on the shop floor has been presented with a mould and it is up to him to make it work.

The only mould modifications which should be recommended in a fault-correction table are to alter the dimensions or geometry of the gate or to arrange for extra mould venting. If a mould refuses to fill properly it is much more likely to be because of insufficient venting than because of a basic fault in design; and if sink marks occur it may well be that the pressure drop in the region of the gate is excessive and can be reduced by some quite minor modification.

A logical and systematic method of dealing with faults is most desirable and many practical moulders have their own scheme of operation. In compiling such a scheme it is important to ensure that all the terms used are unambiguous and are understood by all concerned. Faults should be clearly described and all the possible causes should be examined. The effect of the fault should be taken into account and when the cause has been identified the necessary steps should be taken to eliminate it and to prevent its recurrence. A suggested seven-point scheme is as follows:

(1) *Name the fault.* This may seem obvious but some faults are given a variety of names. For example, 'splash marking' is variously referred to as 'mica marking', 'silvering' or 'silver streaking'. It is advisable to decide which of the names will be used and to stick to that decision.
(2) *Describe the fault.* In describing a fault, a possible cause may be included. For example, a 'short moulding', i.e. an incomplete moulding, may be

described as 'insufficient pressure to fill the mould' or 'not enough material to fill the mould'. It is a useful exercise to try to describe all the common faults in the simplest possible terms without ascribing any possible cause. In this way the mind is cleared of any possible prejudice for the next stage of the scheme, which is finding the cause of the fault.

(3) *Find the cause of the fault*. This may be a lengthy process since it requires consideration of material, machine, mould and process. The following guide lines are suggested:
   (a) *Material*. Check for grade or type, examine for contamination, and make sure it complies with the manufacturer's specification.
   (b) *Machine*. Check the functioning of all parts of the machine in accordance with the recommendations given in Chapter 3, taking into account anything that could influence pressure, temperature, rate, and time.
   (c) *Mould*. Make sure that the mould is properly set and is at the correct temperature, and that all the parts function smoothly and correctly.
   (d) *Process*. Check that the pressure, temperature, rate, and time are set as recommended by the materials supplier.

(4) *Determine the effect of the fault*. If the fault renders the component unusable or unsaleable it must obviously be rectified. If it is only of minor significance it may be unnecessary to try to eliminate it entirely.

(5) *Determine where the responsibility lies*. This may be only of academic interest, but if the fault recurs the operator—as well as the material, machine, mould and process—needs to be checked.

(6) *Take action to avoid the fault*. Failure to take the appropriate action can result in the unnecessary production of faulty mouldings and its consequent detrimental effect on the profitability of the project.

(7) *Take steps to prevent a recurrence*. Make sure that full records are taken of conditions when the fault has been eliminated. Note any repairs and alterations which are made to the mould or the machine and any variations in type, grade or quality of material. If rework (i.e., re-ground scrap) is used, note the proportion and quality.

These steps for dealing with faults may seen superfluous but no fault-finding exercise is complete unless all seven points have been considered. Making faulty mouldings, even though they can all be recovered and re-ground and the material used again, is very uneconomic and is a short route to bankruptcy. Operators making faulty mouldings from whatever cause can have little job satisfaction and if a bonus scheme is operated cannot earn satisfactory wages. Pride in production quickly disappears and the moulding shop personnel assume a *laisser faire* attitude which shows itself in poor housekeeping, poor efficiency and general dissatisfaction.

Dealing with the situation in a positive manner enhances morale, the moulding shop assumes a more business-like aspect, and efficiency increases. The following chapters repeat this message in greater detail.

2

Principles of the process

Injection moulding can be said to have started in 1872 with the filing in November of that year, by I.S. and J. W. Hyatt, of Albany, New York, of US Patent No. 133229 entitled 'Processes and Apparatus for Manufacturing Pyroxiline'. The machine described in the patent specification bears a strong resemblance to the present-day plunger-operated injection moulding machine, having a heated cylinder, a hydraulically operated plunger, a torpedo-shaped piece of metal in the centre of the cylinder to spread material round the heated cylinder walls, and a heated pipe through which the softened material passed into the mould. This was held closed in a hydraulic press referred to in the text of the specification as a 'hydraulic engine'. The cylinder was held in a vertical position and contained material for several moulding shots. Being a rather inflammable material, pyroxiline, mixed with camphor gum to give Celluloid, was warmed very gently so as not to cause it to explode and the pressure was applied slowly. This was to give ample time so that, as the patent states, every portion of the mass of material is . . . 'brought into contact with the heated surface of the cylinder . . . thereby insuring a complete transformation and a homogeneous compound'. The Hyatt Brothers referred to the cylinder as a 'stuffing cylinder' and this has given rise to the common name of 'stuffing machine' for this early form of injection moulding machine.

There seems to have been little development in the process of injection moulding in the interval between 1872 and the mid-1930's. Compression moulding had been developed with the invention in 1908 of Bakelite, a phenol-formaldehyde moulding material, and with the urea-formaldehyde moulding materials of the mid-1920's, but any injection moulding machines of that period tended to be small, hand-operated and to have rather rudimentary mould locking. It was not until the wide range of thermoplastic materials came along just before, during and immediately after the Second World War that injection moulding began to make rapid forward strides.

It was generally thought that the moulding of thermoplastics was very similar to the pressure die-casting of metals and that machines used for that purpose could be adapted or modified in design to mould thermoplastics. This led to the development of machines having a very strong mould locking mechanism and small plasticizing cylinders. The pressure reaching the material in the mould was unpredictable, so to allow for all contingencies it was recommended that a mould lock of 5–10 tons per square inch of the projected area of the moulding should be used. It was not until the middle of the 1950's that the manufacturers of machines and the main raw material manufacturers began to realize that the materials they were dealing with had rather different properties from those of metals and began to think of designing machines to suit them.

Injection moulding, then, is a process whereby a plastic moulding material is fed, in solid form, to a machine where it is heated to soften (or plasticize) it and is then forced under pressure into a mould. In the mould it cools and hardens (in the moulding of thermoplastics) or cures and hardens (in the moulding of thermosetting materials) and in its hardened state can be ejected from the mould which can then be closed and the cycle of operations repeated.

The machine consists of three basic elements:

(1) A *power unit*, generally a hydraulic pressure system in which an electric motor is used to drive a hydraulic pump to produce hydraulic pressure in a fluid which can then be piped to the various sections of the machine to actuate the moving parts.

(2) A *plasticizing unit* for heating and softening the plastic and for forcing the softened material into a mould. This is generally achieved by a plunger, close-fitting to the bore of the cylinder, which is moved forwards and backwards by hydraulic power.

(3) A *split mould* which can be held closed against the pressure of the injected material and opened to remove the resultant moulding.

Most machines are self-contained units but there is no overriding reason why the power system could not be external to the machine and, in fact, why a single power unit could not be made to operate a number of machines. 'Piped' hydraulic power systems serving a number of machines have been tried but in the past have not proved very satisfactory.

The moulding cycle

The operating cycle of an injection moulding machine can best be understood by referring to a diagram such as that shown in Fig. 2.1

(1) *Injection.* Starting with the mould closed (the 12-o'clock position) the first stage of the cycle is injection of material into the mould. For reasons discussed later it is nearly always desirable for injection to take place as rapidly as possible (there are important exceptions, also to be discussed). The machine design—including such matters as the power available and the flow of hydraulic fluid under pressure—determines the minimum time which this will take, but it is possible for the operator to increase the time, that is, to slow down the rate of injection.

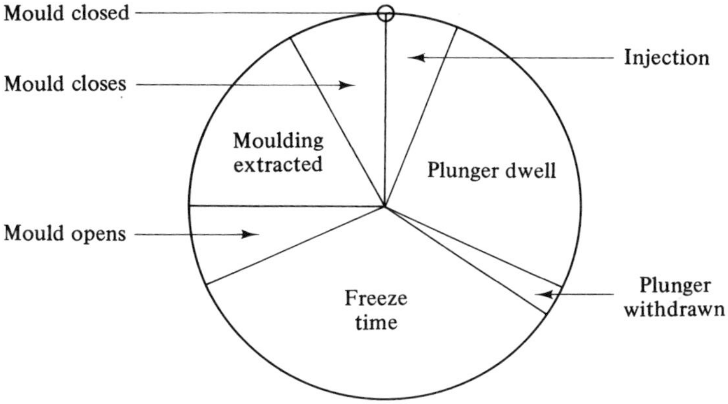

Fig. 2.1 Cycle of operations (plunger machine)

(2) *Plunger dwell.* The next stage is known as 'plunger-dwell time' or 'holding time'. This allows for a little more material to enter the mould before it solidifies in the narrowest portion of the flow channels into the mould (this is generally the 'gate' into the mould cavity) and so prevent material under pressure in the mould from flowing back into the cylinder. The length of time required depends on the properties of the material being moulded.

(3) *Plunger withdrawal and freeze time.* After plunger-dwell the plunger is withdrawn (or the screw moves back) and the next charge of material is fed into the cylinder. If at this stage the moulding had become hard enough to be removed from the mould without distortion no further cooling time would be necessary, but generally a further period known as the 'freeze time' or, as often indicated on early machines, the 'cure time'—a legacy from the thermosetting moulding era—is required. Because material in the mould starts to cool immediately injection has taken place, the true cooling time is the plunger dwell time plus the plunger withdrawal time plus the freeze time. The total time depends largely on the section thickness and can be calculated by means of a formula given in Chapter 4, assuming that there are adequate means of cooling the mould.

(4) *Mould opening and closing and moulding ejection.* The mould then opens and the moulding is ejected, after which the mould closes again ready for the next cycle to begin. Moulds are relatively massive pieces of metal and a considerable amount of power is required to move them. It is desirable both to open and close the mould as expeditiously as possible, but too rapid a movement in either direction tends to put undue strain on the equipment and if the faces of the mould come together with sudden force, this can damage the edges of cavities. Adequate time must be allowed for the ejection of the moulding from the mould either manually or automatically. This time depends on the dimensions of the moulding, which determine the time taken for it to fall free of moving parts between the machine platens and on the

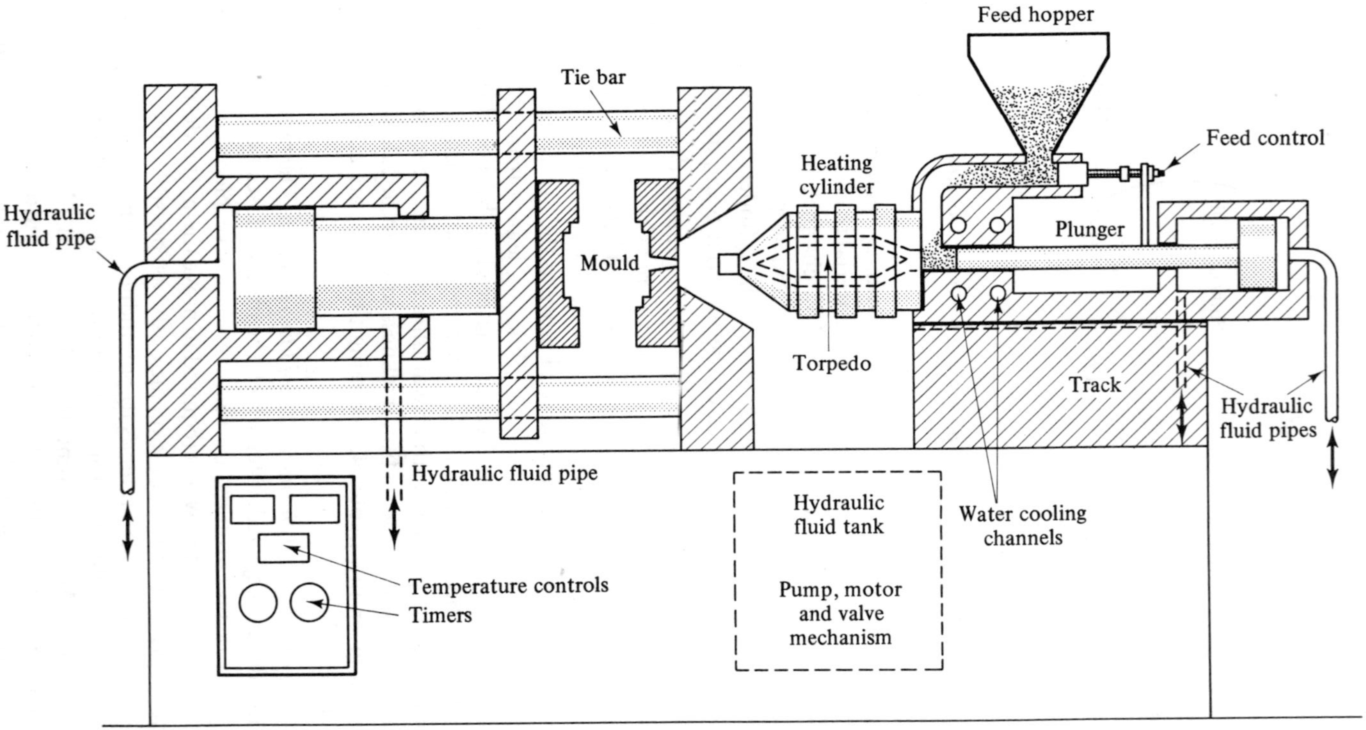

Fig. 2.2 Plunger-operated machine

design and the manufactured quality of the mould. A badly made mould may be a cause of mouldings being difficult to remove from the mould.

From this it will be seen that there are three factors which influence the overall time cycle:

(1) The physical (thermal and rheological) properties of the material combined with the shape of the component.
(2) The design and quality of workmanship in the mould and the design of the mould cooling system.
(3) The design and performance of the machine (and by this is also implied the skill of the setter who fixes the operating cycle).

Most machinery manufacturers quote a 'dry cycle time' for their machines. This is the time estimated or determined from experiment for the machine to operate without any material present, that is, injection, plunger or screw withdrawal, mould open, and mould close.

Plunger machine layout

Consider now the layout of the 'conventional' plunger-operated injection moulding machine of circa 1950 (Fig. 2.2). It consists of a rigid structure, carrying at one end a mould locking mechanism and at the other end a heated cylinder for softening or melting the plastic material. A plunger, actuated by a hydraulic cylinder, moves inside the cylinder to force material into the mould. In the centre of the machine is a fixed platen on to which one part of the mould is clamped and the other mould portion is clamped on a moving platen which slides forward to bring the faces of the mould together and is carried on the tie bars. In the centre of the fixed platen is a hole which registers with a concentric hole into the mould, through which the mould cavities may be filled. The cylinder can slide forwards or backwards on a carriage and the nozzle of the cylinder can be brought forward to register with the hole into the mould cavity. Within the structure of the machine are the power unit—motor, pump, hydraulic fluid tank, cooler unit, etc., and the valves and other machine control equipment.

Plunger machine operation

Material in granular or powder form is fed to the heating cylinder from the feed hopper, the correct amount for each shot being supplied through a self-compensating volumetric metering device. The cylinder contains enough material for several shots and as new material enters the rear of the cylinder the heated, plasticized and fluid plastic is displaced into the mould. In the rear of the cylinder the material is therefore a solid compacted mass of cold granules; at the front, or nozzle, end it is in fluid form, while in between there is a mixture of solid, semi-solid and fluid plastic—a heterogeneous mixture on which the plunger has to act. Inside the main part of the cylinder is the torpedo or spreader, the function of which is to direct the plastic towards the heated cylinder walls in thin streams so that it undergoes the best possible heating effect.

If one then considers the effect of these items on the pressure being exerted on the material which is being injected by the movement of the plunger, it will be seen that there are several reasons why pressure is lost and why it can vary from shot to shot. Unless conditions are absolutely steady, that is, exactly the same amount of material enters the cylinder as leaves it at each shot, there will be variations in the proportions of the solid, semi-solid, and fluid material and consequently variations in the pressure used up in compressing it. These variations will be accentuated by the presence of the torpedo which, in any case, causes a considerable loss of pressure and the net result is that there are considerable, but more important, unpredictable variations in the pressure being applied to material entering the mould. Because of the variations of pressure and their unpredictability it was necessary to allow for a locking force on the mould sufficient to prevent it opening under the worst conditions. That is, to allow for the full pressure of injection reaching the mould and exerting an opening force equal to the pressure multiplied by the projected area of the moulding. Thus, a machine with a maximum injection pressure of 22 000 lb/in^2 (1550 kg/cm^2) moulding an article of area 10 in^2 (64·5 cm^2) would require a lock of 100 tonnes.

Exact metering of the material to the cylinder was also found to present difficulties. Because the metering was on a volume basis, any variation in the bulk density of the material caused alterations in the weight of feed per shot—the mould must, of course, always be supplied with the same quantity of material. Feeding by weight requires expensive and somewhat unreliable equipment and when using some types of material even in a self-compensating volumetric feed device it was difficult to control the volume accurately. Powders tended to flow into the feed throat of the cylinder from the hopper by gravity.

Although at the time it was not fully understood what was taking place the unpredictability of the process was being accentuated because the rheological properties of the plastic melt were being affected by pressure variations. Under some conditions injection would be fast, whilst at times when a good deal of pressure was being lost injection would be much slower. As will be seen in Chapter 4, changes in the rate of injection alter the viscosity characteristics of the molten plastic and this causes corresponding changes in the pressure applied to the material within the mould cavity.

Inherent weaknesses

To attempt to overcome some of these problems the machine manufacturers experimented with a number of systems. It was argued that if the exact amount of *fluid* material could be injected into the mould from a cylinder containing only fully plasticized plastic, predictable mould filling and hence accurate

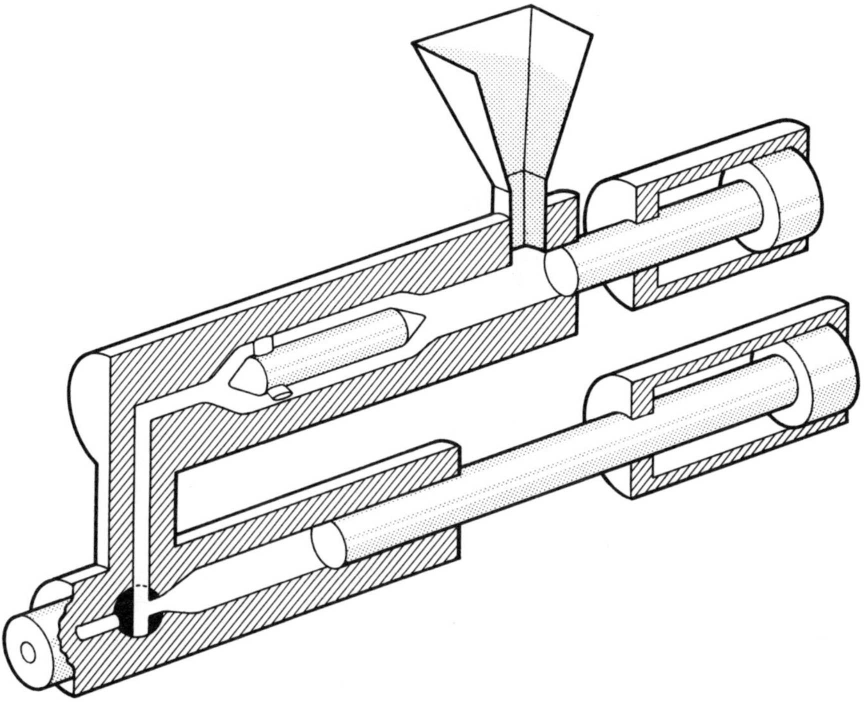

Fig. 2.3 Two-cylinder machine

mould locking could be achieved. The first machine developed had two cylinders in parallel, generally arranged one above the other (Fig. 2.3). The upper cylinder was the conventional plasticizing cylinder containing a torpedo or spreader and received the new charge of solid, granular material. At the front of this cylinder, instead of the nozzle leading to the mould there was a transfer pipe which conveyed material into the lower cylinder. A non-return valve in this pipe prevented flow-back of plastic into the plasticizing cylinder when the lower cylinder was being used for injection. The exact amount of fluid material required for filling the mould could be conveyed to and stored in the lower cylinder by arranging for a limit switch on the lower plunger to arrest the movement of the plunger in the upper cylinder when the lower plunger had moved back far enough.

The sequence of operations was as follows:

(1) Non-return valve opened and material plasticized in the upper cylinder entered the lower cylinder.
(2) Lower plunger moved back and operated switch to stop upper plunger movement.
(3) Non-return valve closed and lower plunger moved forward to inject material into mould.

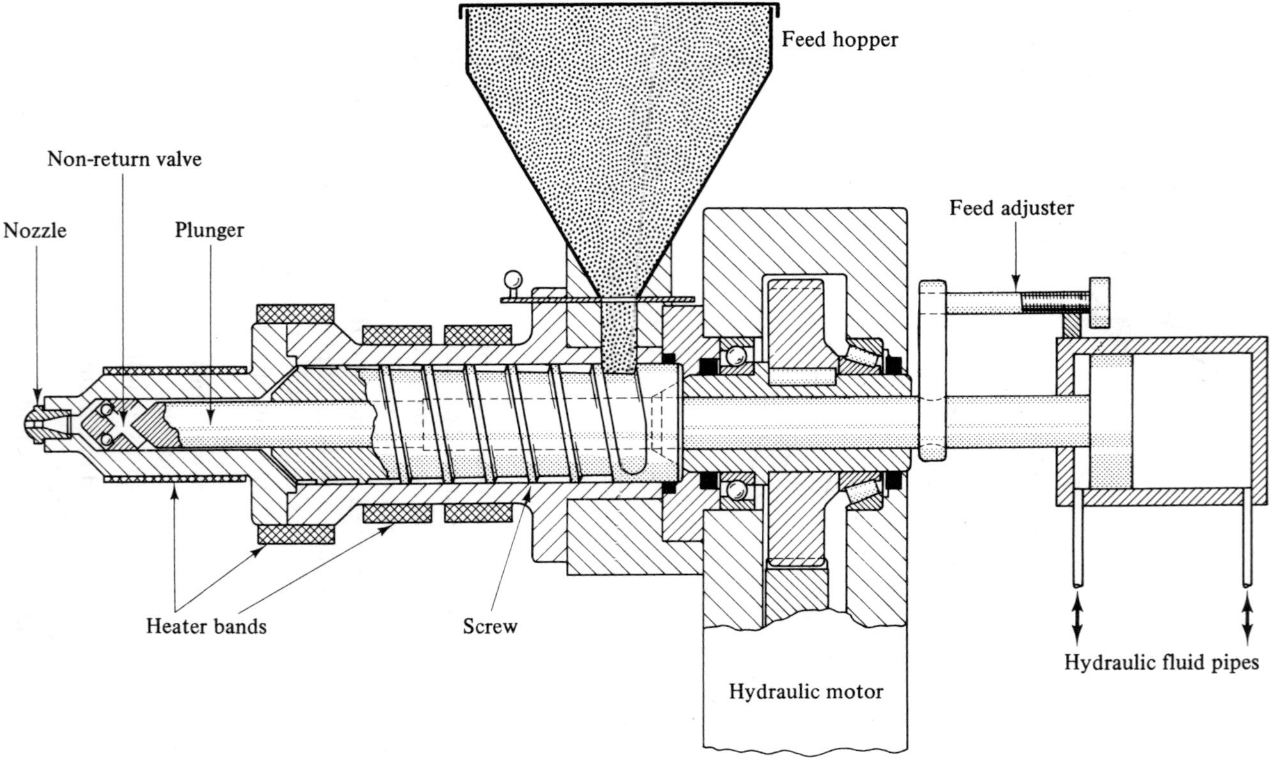

Fig. 2.4 Co-axial screw machine

(4) Upper plunger moved back and a new charge of material entered upper cylinder.
(5) Non-return valve opened and upper plunger moved forward to recharge the lower cylinder.
(6) Mould opened to eject the moulding.
(7) Mould closed ready for next shot.
(8) Injection, and stages (3) to (7) repeated.

At close-down it was advisable to empty both upper and lower cylinders and, as far as possible, the transfer pipe to reduce the difficulty of moving a semi-solid mass of material in the pipe and in the lower cylinder without causing overheating and consequent decomposition when re-starting the machine.

The main problems associated with the use and operation of the twin parallel cylinder machine were:

(a) A considerably more complicated and expensive electrical switching system than previously used was necessary.
(b) It was difficult to maintain the material in the transfer pipe at the correct degree of fluidity without causing overheating and decomposition.
(c) To keep the machine size and the length of the transfer pipe to a minimum it was necessary to arrange the machine so that the new fluid material entered the injection cylinder at the front end and there tended to be a residue of hot material staying at the head of the injection plunger. (Later versions of the machine were redesigned to allow the injection plunger to 'sweep' the inlet port and this largely obviated the trouble.)

Other attempts were made to design two-cylinder pre-plasticization machines and included in-line twin cylinders, vertical accumulator systems similar to those now used in extrusion blow-moulding, and the combination of screw-plasticization units such as are used in extrusion, with a standard plunger injection cylinder. The in-line twin cylinder machine generally had a fairly standard plasticization cylinder which was used to charge what might be termed an over-size nozzle section from which an extended plunger injected the material into the mould. Another system, known as the co-axial screw machine (Fig. 2.4) used a short large diameter screw to plasticize the material and to force the fluid plastic along channels into an enlarged nozzle section. The screw was bored along its length, enabling a long plunger to reciprocate through it, actuated by means of a hydraulic cylinder at the rear of the machine, and operating to produce injection by 'sweeping' the nozzle section. A non-return valve in the nose of the plunger prevented back-flow into the screw during injection.

The use of an extrusion-type cylinder to plasticize material and then feeding it to an injection cylinder suffered from some of the disadvantages of the twin parallel cylinder machine already described. A fairly short cylinder was used,

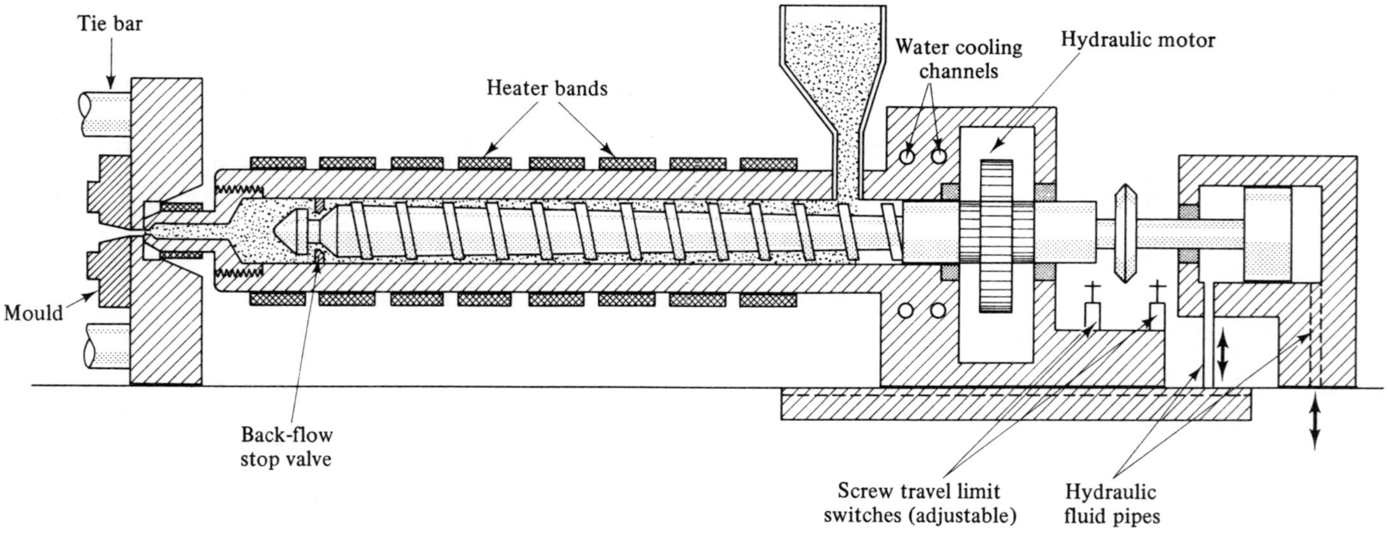

Fig. 2.5 Reciprocating screw machine

often set at an angle of 30–40° to the horizontal and feeding into the front of the injection cylinder. A non-return valve prevented material feeding back into the screw during injection. The system gave somewhat better plasticization than the plunger-type plasticizing unit and a more homogeneous melt. However, control could be difficult, especially when using the more heat-sensitive plastics.

In spite of the many drawbacks of the plunger machine, it was not until the arrival of the reciprocating screw machine that it was finally superseded. Even so, some moulders in the United States of America continued to operate the larger plunger machines for some considerable time after the reciprocating screw machine had become standard in Europe. The very small machines used nowadays chiefly for prototype work continue to use the plunger principle.

Reciprocating screw machine

Experiments in the 1930's had shown that it was possible to use a rotating screw to plasticize material for injection moulding and then to transfer the material into the mould by pushing forward the screw to act as a plunger. At that time, with only a limited range of thermoplastic materials available and a flourishing compression moulding industry for thermosetting plastics, the injection moulding side of the business was not ready for innovation. It was only in the mid-1950's that the re-discovery of the idea could be exploited and enable a great step forward to take place in the technological aspects of injection moulding.

The layout of a typical reciprocating screw injection moulding machine is shown diagrammatically in Fig. 2.5. All the parts used in the plunger machine described earlier are employed except that, instead of a plunger and torpedo in the cylinder there is an extrusion-type screw. This can be made to rotate by means of an electric or a hydraulic motor and it can be made to move backwards or forwards by means of a hydraulic cylinder such as is used on the plunger machine. Back-flow of material along the screw flights during injection is prevented by a non-return valve (referred to as a back-flow stop valve or check ring) on the nose of the screw. Two types of non-return valve are shown in Fig. 2.6a and 2.6b. The self-compensating feed device is not necessary because the forward and backward motion of the screw is controlled

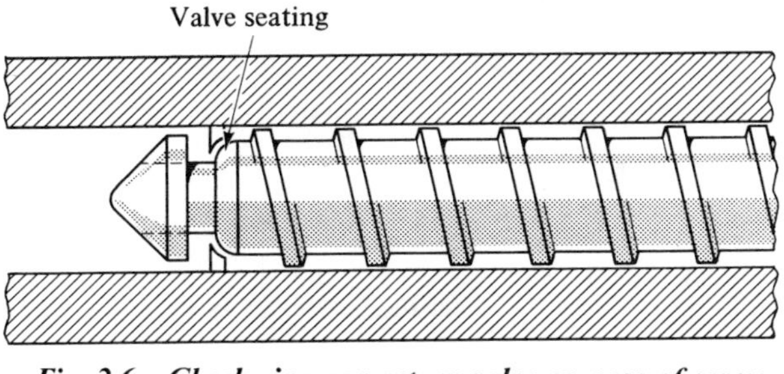

Fig. 2.6a Check ring non-return valve on nose of screw

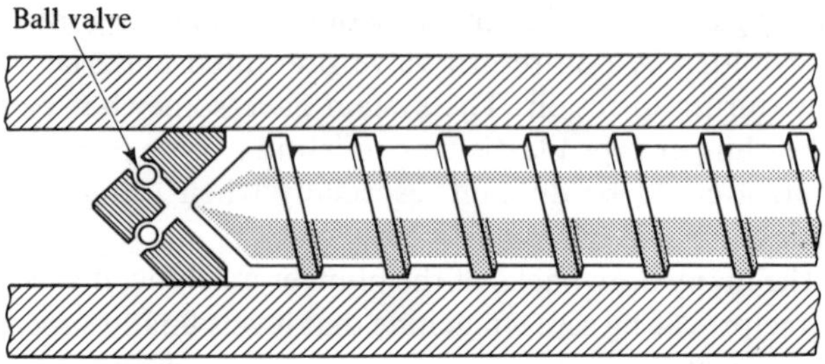

Fig. 2.6b Ball-type non-return valve on nose of screw

by limit switches which act on electrically operated valves to stop screw rotation or to shut off the flow of hydraulic fluid to the cylinder when the injection movement of the screw is complete.

Screw machine operation. A typical sequence of operations from start up is as follows (Fig. 2.7)

(1) Starting with an empty cylinder, granular material from the feed hopper falls on to the rear flights of the screw which rotates, carrying material to the front of the cylinder. During its passage along the cylinder it is plasticized to a fluid state. Some material may escape through the nozzle but the back pressure is generally sufficient to push the screw back in the cylinder and to provide a reservoir of fluid plastic in the front of the cylinder ready for injection.

(2) The mould closes and the cylinder moves forward on its carriage until the nozzle is in contact with the entrance to the mould.

(3) The screw is moved forward by the hydraulic cylinder at the rear of the machine and injection takes place.

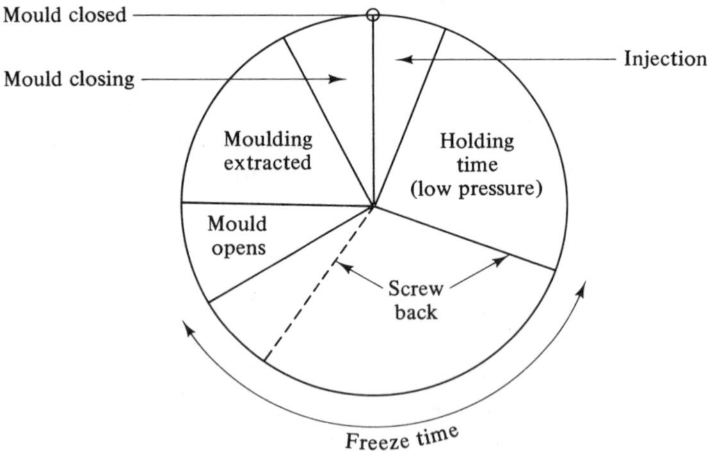

Fig. 2.7 Cycle of operations (screw machine)

(4) After a short interval (the holding time) the screw rotates, creating some pressure in the barrel, which forces it back against low pressure in the hydraulic cylinder, until the limit switch operates, stopping the rotation. This plasticizes material ready for the next shot.
(5) The mould opens, the moulding is ejected, and the mould closes again ready for the next cycle.
(6) Stages (2) to (5) repeat.

Advantages of reciprocating screw machines. There are several advantages to be gained in using the reciprocating screw system. Among these are:

(a) The exact volume of *fluid* material required for filling the mould can be very accurately metered because the limit switches are capable of being adjusted with micrometer precision.
(b) Pressure drop caused by the torpedo is avoided.
(c) It is not necessary to consider weight feeding and the use of complicated self-compensating feed devices is avoided.
(d) The screw is acting as a plunger on a homogeneous melt and the pressure applied to the mould becomes predictable.
(e) Frictional losses in the cylinder are minimal.
(f) Heat transfer in the cylinder is improved because of the mixing action of the rotating screw. This also supplies some mechanical heat because of its shearing action on the viscous plastic.
(g) Because the process is more predictable it is possible to use a smaller locking force for a given area of moulding.
(h) Higher injection speeds are possible.

Figures 2.1 and 2.7 (the cycle of operations) show that the only difference between the cycle for a plunger machine and that for a reciprocating screw machine is the feeding of the new charge to the cylinder. Instead of 'plunger withdrawal', the screw rotates and is displaced backwards along the cylinder, the operation being known as 'screw back'. This takes place during the extension of cooling time (freeze time) but for most efficient operation should not be allowed to extend the freeze time beyond what is actually required for the hardening of the moulding. Most machines are sequenced so that the mould cannot open until screw rotation is complete. This effectively prevents the full cycle operating if the feed hopper becomes empty and also prevents any leakage of material from the nozzle when the mould is opened unless pressure has been developed inside the cylinder during screw back. To obviate any possibility of nozzle leakage because of this, many machines are now fitted with decompression in which, after screw back is complete, the screw is slightly drawn back to relieve the pressure. Decompression—commonly called 'suck back'—should not be excessive or air will be drawn into the hot nozzle where it can cause oxidation of the molten plastic. It is also desirable to keep the screw-back time and the speed of rotation of the screw constant from shot to shot so that the same amount of mechanical heat is supplied to the material at each cycle.

Latest trends and developments in machines

With the arrival of the reciprocating screw machine the technology of injection moulding made rapid advances in two directions. Firstly, in the size of machine possible and secondly, in the better understanding of the problems of material flow in the cylinder and in the mould. It also enabled materials to be injection moulded which had not been possible before because the better and quicker heat transfer accompanied by mixing during plasticization meant that lower cylinder temperatures could be used. Materials of higher melt viscosity could also be moulded because the pressure drop in the cylinder was lower.

Until about 1960 the size of injection moulding machines was limited by two factors:

(1) The relatively low rate of heat transfer through the cylinder walls into the material in the static state of the plastic with plunger type operation.
(2) The excessively large mould locking units which were necessary to ensure that mould flashing would not occur, whatever the conditions.

Many attempts were made to overcome these difficulties, for example, by using rotating torpedos, modifying torpedo design, and by making longer and bigger cylinders. Injection pressures were increased and plasticization took place under pressure. Smaller machines were fitted with 'double shooting' devices so that larger mouldings could be made but, in spite of this, machine sizes in relation to shot weights became impossibly large. Even so, the maximum shot weight was not much more than 3–4 kg.

The advent of the reciprocating screw machine meant that more material could be plasticized in the same size of cylinder. Taking away the torpedo left more room in the cylinder for material and reduced frictional losses. At the same time, because the mould locking force could be lower, larger area mouldings became possible and a 'norm' of 2–$2\frac{1}{2}$ tonnes/in^2 of moulded area was adopted. By 1970 machines capable of moulding 30 kg at one shot were being made in Japan. By an adroit combination of shot volume control, rapid injection, and special mould design, mouldings up to 4 ft square (120 cm square) with an area of 2300 in^2 (14 400 cm^2), and a weight of 22 lb (10 kg) were being moulded in the United Kingdom and it was said that the locking force was no more than 1300 tonnes or just over $\frac{1}{2}$ tonne/in^2 (0·1 tonne/cm^2). For safe working with a plunger machine—had the exercise been possible—a mould locking force of 20 000 tonnes would have been advisable.

Using screw machines, it has been possible to reduce the mould locking force, especially of the smaller models, for a given shot volume (or to increase the shot size for a given locking force) and this has helped to hold down the cost of machines. In an inflationary situation this may not, at first sight, be apparent but it is probably fair to say that, had this technological advance not come along, prices would be much greater than they are in a more restricted market. Typical machines are illustrated in Figs. 2.8 and 2.9.

Fig. 2.8 Modern British machine with hydraulic lock (180 tonne)

Understanding the process

The advance in the understanding of the process has also been in two main directions, both of which are now making possible a much more scientific and mathematically calculable approach to problems, rather than the trial and error methods of two decades ago.

To appreciate the nature of the problems involved in making this scientific approach to injection moulding, it is necessary to study the phenomena of heat transfer through the materials used when they are in process of being moulded—that is, in an unsteady state with regard to enthalpy. This requires information on the thermal properties of the plastics, such as latent heat of fusion, specific heat, and thermal conductivity. These properties must also be considered in association with changes of pressure—often excessive changes—and in the crystalline or fine structure of the plastics. Some of these matters are discussed in Chapter 4 when the properties of materials for injection moulding are being considered, but the outcome of studies that have been made has been to stress the need for considering cylinder heating and mould cooling in much greater detail than in the past. As a result, time cycles and cooling requirements can now be calculated with a fair degree of precision.

Plastic materials in the fluid state do not behave in the normal manner of viscous fluids, such as oil or glycerine. They exhibit a large degree of pseudoplasticity[1] when subjected—as in the injection moulding process—to high rates of shear. This means that the approach to injection moulding which compared the process to the pressure die-casting of metals does not apply except in a limited way, as, for example, in the moulding of very low viscosity materials such as nylon.

The development of the reciprocating screw injection moulding machine with its much more predictable performance required information about the

[1] Pseudoplasticity is the term applied to the flow behaviour of fluids of which the viscosity decreases as shear rate increases.

Fig. 2.9 Modern Italian machine with toggle lock (170 tonne)

flow properties of plastic melts which had not, in the past, been available. To calculate pressure and flow requirements it is necessary to have information on the viscosity characteristics of materials, both at a range of temperatures and at a range of shear rates. An attempt to obtain such information had already been made by the investigation of melt flow index (MFI), whereby the amount of material being extruded through a given orifice in a given time at set temperatures and pressures was determined. This test, however, gives the information only for the set of conditions under which the experiment is conducted and does not relate to different temperatures and rates of shear. Another, more practical approach so far as injection moulding was concerned, involved moulding material into an open channel and measuring the distance of penetration at different temperatures and pressures. (For convenience the channel was in the form of a spiral and the experimental procedure was known as 'spiral flow moulding'.) This technique certainly went a little further towards obtaining results from which calculations could be made, but in the main the information obtained was only of a comparative nature as between different materials or different grades of material.

During the decade 1960–1970 researchers had been investigating the rheological or flow properties of plastic melts in a relatively new piece of equipment known as a high shear rate viscometer. Data had been obtained at the shear rates operating in extrusion cylinders and dies and by extrapolation this could be extended to the shear rates occurring in various parts of the injection moulding system, particularly the entry orifice into the mould and between the faces of the cavity. Much of the information and the conclusions drawn from it must still be considered tentative, but a start has been made. It is now possible to calculate with a fair degree of accuracy the pressure losses between various points in the flow paths of the injection moulding system. Pressure requirements can be calculated and the time for correct injection determined. It is also possible to calculate a suitable starting size for the gate—the restricted point of entry of material into the mould which is so important in determining the quality of the moulded article.

These developments have enabled more accurate diagnoses to be made in fault-finding and trouble-shooting. Many faults and breakdowns occur during the service life of injection mouldings because of the incorrect combination of material properties, mould characteristics, and machine performance during manufacture. By an analytical approach, the use of elementary mathematical calculations and reference to material properties and machinery data it is often possible to determine the cause of failure and to propose steps for rectifying the fault.

Developments in control systems

Another most important result of the investigation of the thermal and rheological properties of plastics, coupled with the development of the reciprocating screw machine, has been the movement towards computerized control of the process. It is now possible, by measuring the temperature of the melted plastic inside the cylinder as it is ready for injection and feeding the information into control equipment, not only to adjust the power input to the cylinder heaters to suit the required conditions, but also to regulate the hydraulic pressure applied to the screw/plunger according to the variation with temperature of the viscosity of the melt. Sensing the pressure within the mould cavity during filling enables the injection pressure and rate to be varied and the shot volume to be corrected. Some of these developments were discussed at a Conference entitled 'Advances in Injection Moulding' arranged by the Plastics and Rubber Institute on 11 and 12 November 1975.

The net results of these and other developments are claimed to be:

(1) Closer dimensional tolerance of mouldings due to exactly metered shot volumes.
(2) Better quality mouldings (free from stress and from areas of highly packed material) and the ability to use previously processed material without a deterioration in quality.
(3) Improvement in output because it is possible to optimize conditions more easily.
(4) Reduction in power requirements, better machine utilization, and down-rating of the mould locking force with relation to the moulding area.

These matters are discussed at greater length in Chapter 4 on materials and Chapter 6 on process control.

Although computer control would appear to be the ultimate answer to problems in injection moulding it will probably continue to be an expensive addition to standard moulding equipment. Machines having computer control built-in will always be more expensive than less sophisticated equipment and may be beyond the reach of the smaller company.

A machine that is well maintained with instruments and control equipment kept in good condition and tested from time to time will always perform better than badly maintained, dirty equipment which so often has to serve. Even an old machine, provided its essential working parts are in good running order, can be made to produce good work in the hands of someone who thoroughly understands its operation and is familiar with the basic theoretical information on the subject. The computer may, indeed, take away some of the skill of the operator but ultimately it is the profitability of a process which determines the degree of sophistication of control which can be applied. If, by using a properly trained operator an equal degree of profitability can be achieved as with a computer-controlled machine and at the same cost, there would appear to be little argument for change.

3

Machines

Although injection moulding machines come in a variety of shapes and sizes, they all consist of three basic parts (Fig. 2.2 and 2.4): the power unit, the plasticizing unit and the mould unit. The arrangement of these units need not necessarily be as shown in the diagrams. For example, the mould unit may lock in the vertical plane rather than the horizontal; the plasticizing unit may be disposed vertically rather than horizontally and the power unit may be quite separate from the main machine. However, the general tendency is to make each machine self-contained so that, by connecting to the electric power supply and to a suitable cooling water system—with, perhaps, a compressed air supply—it will operate independently of any other machine in the factory.

There is little doubt that economies could be effected by a return to the system common in the compression moulding shops of 20 or more years ago, where a central hydraulic system supplied a number of machines. There are both advantages and disadvantages in such a system, quite apart from the fact that no manufacturer is at present producing suitable machines. Hydraulic lines and supplies would need to be of sufficient size and capacity to allow simultaneous operation of all machines and duplication would be necessary in case of leakage or breakdown. These and other problems would need to be set against the resulting quietness in the moulding shop and the important saving of electric power.

The power unit

(1) *The hydraulic pump.* The heart of the injection moulding machine is the hydraulic pump and the electric motor which drives it. In most machines the pump is of the rotary variety and, close coupled with the motor, makes a compact unit. The smaller machines use one motor only, but larger machines have two, three, or more motors and pumps, some supplying hydraulic fluid at low pressure and others of smaller capacity giving the high pressures necessary for final mould locking, injection, and screw drive. Of the many types of rotary hydraulic pump the most common is the vane pump.

The simplest form of vane pump consists of a solid rotor having radial slots into which slide closely fitting vanes. Surrounding the rotor assembly is an eccentrically fitted ring in which it rotates, carried inside a casing with end plates having bearings and seals for the rotor. A series of inlet and outlet ports is arranged at each side of the ring. When the pump is in operation, the vanes in the rotor are thrown outwards by centrifugal force so that their tips sweep the inside surface of the ring. As they pass the inlet ports the volume enclosed by two adjacent vanes is increasing and drawing in fluid, but as the same two vanes pass the outlet port the volume is decreasing and so increasing the pressure on the fluid. This type of pump delivers a fixed quantity of fluid which is determined solely by the speed of revolution, but by altering the position of the ring in relation to the rotor the volume can be adjusted. It is possible to alter the position by means of a plunger and spring tensioner and this can be arranged to function automatically according to the pressure and flow required.

Pumps of this simple type cannot be balanced and the bearings tend to wear out more quickly on one side than the other. To obviate this, the ring is generally given a cam form with two lobes and two sets of inlet and outlet ports. It is then more complicated to adjust the relative position of the cam ring and rotor so other means of controlling the pressure need to be used. If the pump were allowed to operate in a closed circuit, pressure would build up until the pump casing burst. A pressure relief valve which will open at a predetermined pressure and allow fluid to flow back to the supply tank is essential for the operation of this type of pump.

Rotary vane pumps rely for their continuous working on the fact that the hydraulic fluid which is being pressurized acts as a lubricant. Should the pump become starved of fluid for any reason, such as a blockage in the supply to the inlet port, the pump is said to 'cavitate'. Lubrication is lost and the cavitation quickly causes erosion of the cam ring, the vanes become worn, and the pump rapidly deteriorates, losing its ability to pressurize the fluid and finally breaking down.

(2) *Hydraulic fluid reservoir*. To allow the hydraulic pump to function, a reservoir of hydraulic fluid is required. In many small machines this is accommodated in the base, but with some machine arrangements it is more convenient to have the reservoir elevated above the machine. The advantages and disadvantages of each arrangement will become obvious during discussions in later chapters. In any event, it is important to have the reservoir, as far as is possible, totally enclosed to prevent dust and dirt getting in. As will become apparent when discussing faults in mouldings, contamination is the greatest enemy of high-quality production techniques.

In an enclosed tank there must be a filler cap which can also act as a 'breather' to allow air to enter or leave the reservoir as hydraulic fluid leaves or enters. A filter gauze in the cap will keep out airborne dust and dirt. The tank needs to be of large enough capacity to contain enough fluid to supply all the machine functions without becoming exhausted and to hold all the fluid which may drain back when the machine is at rest, plus a little spare capacity.

(3) *Hydraulic fluid contamination.* Some machines have a submerged pump housed within the reservoir and sealed from the motor drive. More commonly, a wide-bore pipe on the end of which, inside the tank, is a fairly coarse strainer, supplies fluid by gravity to the pump. The inlet to this pipe, and to the pump if submerged in the reservoir, should be a few inches (5–10 cm) above the bottom of the tank. This allows for any sludge or residue which may accumulate to fall undisturbed to the bottom of the tank and not to be drawn into the pump where it might cause a blockage on the inlet side. In some machines, should the strainer become blocked, a partial vacuum switch operates to cut out the pump motor and prevent any damage to the vanes by lack of hydraulic fluid.

Also in the tank, or in a compartment in the fluid supply line, is an assembly of permanent magnets which attracts any particles of magnetic ferrous metals, so helping to reduce excessive wear on the pump. A very fine filter is usually incorporated in the fluid line, on the outlet side of the pump, to remove any solid particles which could cause wear on the moving parts of the pump or block any other parts of the hydraulic system.

(4) *Hydraulic fluid temperature.* During the pumping process the hydraulic fluid becomes warm and needs to be cooled. Some machine manufacturers favour an internal cooler in the reservoir which consists of a copper tube, coiled appropriately, through which cold water is passed. Most modern machines have external heat exchangers installed in the hydraulic fluid line away from the reservoir so that they can easily be isolated and removed from the system for cleaning. It is desirable to have a full flow of water through the heat exchanger at a temperature which does not cause too much chilling. If the fluid in contact with the cooling tube becomes too cold its viscosity increases so much that it does not transfer the heat efficiently. In the worst conditions the bulk of the fluid may hardly be cooled at all because the layer of sluggishly moving material insulates the cooling tubes. A state could be envisaged where portions of viscous fluid could be detached from the cooling tubes and float about in the remainder of the oil, causing variations in the pressure and flow to various parts of the system. On the other hand, if the temperature gets too high, it may cause the fluid to lose so much viscosity that a loss of pressure results. Many machines have a thermal switch which operates at a temperature of about 55°C, causing a valve to open on the outlet side of the pump, and reducing the pressure and returning the fluid to the reservoir. As soon as the temperature goes down the valve closes and normal machine functioning resumes.

(5) *Accumulators.* The demand for hydraulic fluid does not stay constant throughout the moulding cycle and for this reason it is desirable to store some of the energy generated during slack periods so that it can be used when demand is high. A hydraulic accumulator can be used for this purpose.

In its simplest form, a hydraulic accumulator consists of a vertical pipe of appropriate length and diameter to contain the desired volume of fluid. The lower end is connected to the pump through a system of valves and above the upper surface of the fluid in the pipe is a close fitting plunger which will slide up or down the tube. On the upper side of the plunger is a weight and as fluid is pumped into the pipe the plunger and weight rise. Valves and switches control the movement. When hydraulic fluid is required it can be drawn off and the level of the plunger falls. The weight applied to the plunger and the diameter of the pipe are calculated to suit requirements. For example, an 8 in (200 mm) diameter pipe with a 50 tonne weight will give a line pressure of 2000 lb/in^2 (140 kg/cm^2) and will have a capacity of approximately $2\frac{1}{4}$ gal/ft (31·4 l/m) length of pipe.

The more commonly used hydraulic accumulator used in injection moulding machines consists of a high-pressure cylinder of appropriate size in which is a strong rubber bag containing nitrogen at a pressure of about 1000 lb/in^2 (70 kg/cm^2). Hydraulic oil can be pumped into the cylinder against the gas pressure and stored under pressure until required. Accumulators of this type are supplied in various sizes to suit the capacity of the machines.

By using a hydraulic accumulator it is possible to reduce the capacity of the pump and motor and yet to ensure that injection can take place at whatever speed and pressure is required. The saving in cost of the pump and motor will easily pay for the cost of the accumulator, including the valve and switching mechanism. If a machine is not able to perform as desired because the pump is too small, a hydraulic accumulator will make it possible to up-rate the performance. Specialist firms advertise their services and can be called in to fit accumulators if desired, but many machines now have them as standard equipment.

Hydraulic fluid

If the pump is the heart of the injection moulding machine, the hydraulic fluid is its life blood. It is often referred to as 'hydraulic oil' or simply, 'oil'. Hydraulic fluid is generally a specially formulated hydrocarbon oil to which is added a viscosity depressant which has the effect of causing the viscosity of the oil to remain constant over a wide temperature range. The viscosity is virtually independent of pressure. Hydrocarbon oils are all more or less inflammable and to guard against fires some moulders are anxious to use non-inflammable oils. Some of these are based on hydrocarbons having flame-retardent additives and others on organic phosphates, such as tri-xylenyl phosphate. Most of them are considerably more expensive than straight hydrocarbon oils and it is a matter for debate whether it is better to ensure safety by using non-inflammable oils or to work to a higher standard of machinery maintenance.

Like all hydrocarbons, hydraulic fluid is subject to the effects of temperature and shear. It is possible for the oil to break down into simpler, reactive compounds and then to re-combine into a more complex form, the worst possible effect being to change the fairly viscous fluid into a mixture of jelly-like material suspended in a very low viscosity medium. Such a change would, of course, take a very long time, but the effects of having such a fluid in a hydraulic system can be imagined. Another possibility is for the oil to become oxidized, for its viscosity to increase, particularly at low temperatures, and for its performance as a hydraulic fluid to be almost unpredictable.

Normally, oil and water do not mix, but under the shearing effect of some control valves it is possible for any water droplets to become emulsified in the oil with a detrimental effect on performance. Air, too, which becomes entrained in the oil, can become emulsified and then will not separate out. Both air and water emulsions in oil have much reduced lubricating properties compared with pure oil; this can be detrimental to the pump, but both contaminants also affect the compressibility of the oil which can lead to variations in pressure and rates of flow, giving shot weight variations in the moulded components.

It is therefore essential to prevent air and water from entering the hydraulic fluid system. The former can be prevented by ensuring that a partial vacuum never occurs in the inlet to the pump and by bleeding out any air which enters pipes which have been removed for maintenance. The ingress of moisture can be reduced by ensuring that the temperature of the oil during machine operation never falls below the dew-point of the surrounding air. Moist air entering a cold reservoir to displace oil that has been pumped into the system can easily deposit quite considerable amounts of water simply by condensation.

The manufacturers of machines and of hydraulic oils alike all recommend that the oil should be changed at regular intervals. A period of 3000 h of running time is often suggested, but if reasonable care is taken this may be greatly extended. It is not, however, good practice to continue to operate with oil which is obviously in poor condition; poor quality mouldings and indifferent machine control will result and the net losses incurred may be greater than the cost of a new supply of hydraulic fluid.

Running temperatures in excess of about 55°C are not recommended and it is good practice not to have the oil temperature lower than 25°C. In fact, machines function most satisfactorily if the oil is kept at 35°C and if (for example, after a winter shut-down) the oil should get really cold, it is advisable to heat it to at least 10–15°C before trying to run the machine. At temperatures near to 0°C the viscosity of old oil may be so great as to cause cavitation in the pump with possible risk of breakdown.

Because the hydraulic oil is so important for the efficient running of the moulding machine it is worth while to take the greatest possible care to avoid contamination by conditions other than those recommended by the manufacturers. Its high cost alone should be a factor in this aspect of good practice. However, it is to be regretted that many moulders do not pay enough attention to the hydraulic oil and it has even been known, when leaks have developed in a pipeline, for the oil to be collected in an old can or bucket and then poured

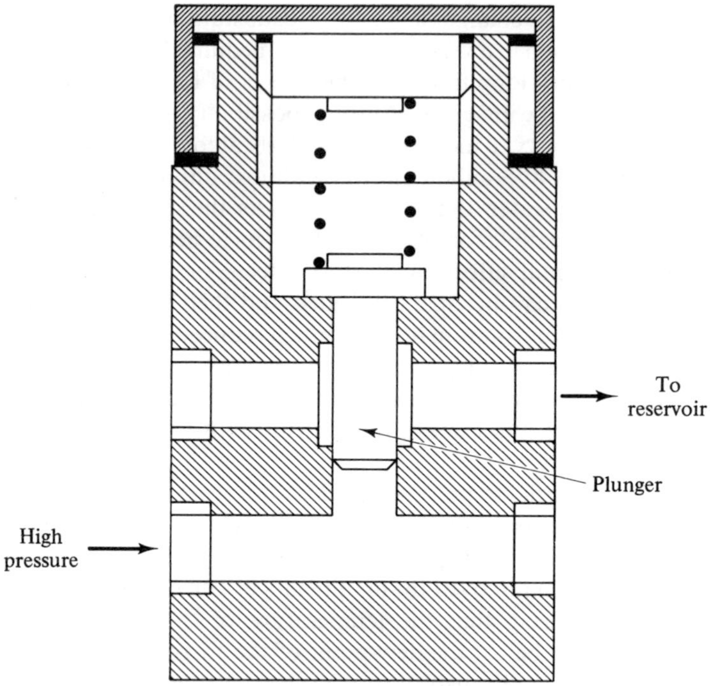

Fig. 3.1 Simple pressure regulating valve

back—dirt and all—into the tank. Many faults in injection moulding, particularly those involving shot-weight variation, can be fairly attributed to lack of care for the life blood of the machine.

Hydraulic oil control

As has been previously indicated, to prevent the oil pressure from rising to an undesirable level, a pressure regulating valve is incorporated in the circuit. Figure 3.1 shows the working of a simple pressure regulating valve and Fig. 3.2 shows a schematic circuit arrangement. When the pressure reaches a predetermined level, the plunger, acting against the spring, is forced upwards, causing the valve to open and to release oil to the reservoir. It will readily be appreciated that any particles of dirt lodging under the valve seating could completely upset the pressure control of the system. If the valve is not of sound design and robust construction it is possible for it to vibrate and cause a great deal of noise as well as being less than positive in its action. As the oil passes through the needle valve it is sheared and becomes heated, some of the heat being transferred to the valve itself. The resultant expansion of the metal can alter the characteristics of the valve so that it needs to be adjusted from time to time, using the screw to vary the pressure on the spring. This variation can affect the quality of the mouldings (discussed in Chapter 6).

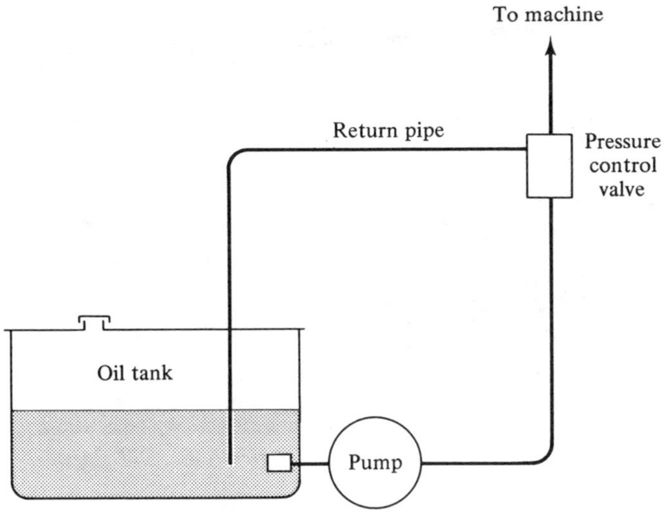

Fig. 3.2 Schematic circuit for pressure control

Hydraulic pipework

The hydraulic oil is conveyed to the various parts of the machine by means of pipelines. Where possible these are of metal, but because many of the parts of the machine move relative to the base framework, a large proportion of the pipework has to be flexible. Although the flexible pipes are nowadays extremely reliable they are obviously a little less robust than the metal pipes and need to be inspected regularly for signs of failure. One important aspect of flexible pipework is the manner in which it is connected to the rigid pipes. If the connection brings the flexible pipe out in a horizontal direction there is liable to be a considerable strain at the point of connection when the part to which one end is fixed moves horizontally with relation to the other end. It is always desirable to have flexible pipes connected vertically to parts which move horizontally so that the natural bend of the pipe takes up any strain (Fig. 3.3). If connections *must* be horizontal, the flexible pipe should be supported for a short distance away from the joint (Fig. 3.4). Failures of flexible pipelines are fortunately rare but if they do occur they can be dangerous. Hydraulic oil under high pressure escaping through a small slit becomes atomized and therefore highly explosive when mixed with air. Furthermore, if a flexible pipe becomes completely detached it does not spray the oil out in a single direction but snakes about drenching the surroundings until the pressure is exhausted. Rapid action to stop the motor is necessary if this occurs.

Machine sequencing

Solenoid valves. Sequencing and actuating of the various parts of the machine are brought about by the opening and closing of valves, using

Fig. 3.3 Flexible pipes connected vertically to rigid pipework

Fig. 3.4 Horizontally connected flexible pipe clamped to restrict movement at joint

electrically operated solenoids. Such valves are called 'solenoid valves'. The power supply to the solenoids is controlled by electric relays which switch the high amperages required for the solenoids and are themselves operated by low voltage and wattage currents switched on and off by means of microswitches on the machine. The solenoid switching may also be effected using solid-state electronic circuitry. Because both relays and solenoids have moving parts, their operation, though rapid, does take a finite time and the heavier the duty being performed, the longer is the time.

Solenoid valves are suitable for handling flows of up to 3 gal per minute (240 cm^3/s). This is satisfactory for most small machines up to about 4 oz (120 g) capacity but for larger machines a more complex valve is required. In this, the solenoid actuates a pilot valve which supplies hydraulic fluid to the spool of a large valve, causing it to move to open or close, as required. The actuating time for a valve of this type is longer than for the smaller valve and is of significance in some aspects of the injection moulding cycle, for example, in switching to low-pressure hold after high-pressure injection.

Microswitches. Many types of microswitch are in use for switching relays and solenoids. They all have moving contacts which require to be kept extremely clean. In common with the relays and solenoids, they are generally housed in dust-proof cases with rubber seals or gaiters to prevent dust from entering at moving parts. If the seal or gaiter becomes perished, punctured, or displaced, allowing dust to enter, the operation of the mechanism is slowed and may fail altogether. Although solenoids and relays are normally housed in dust-proof cabinets, it is distressing to see the number of instances where cabinets are left with doors open and panels removed with, perhaps, a granulator operating and producing large volumes of dust a few inches away.

Many microswitches on injection moulding machines are actuated by cams, the microswitch being equipped with a cam follower which may be a plastic or metal wheel. It is virtually impossible to prevent dust settling on the cam and wheel, and here again, the setting of the switch and the actuating time can be affected adversely. Instead of microswitches, it is sometimes possible to use photoelectric relay controls or electromagnetic proximity switches. In the former, the need for cleanliness will be obvious. With the latter, the switch electrically senses its proximity to a piece of steel or a small magnet, but even with this type of switch, the presence of plastic dust with a high permittivity value can upset the position at which sensing takes place. However, with both of these types of sensing device, which have no moving parts to become clogged with dust, one source of delay in the actuating time is eliminated.

Electronic circuitry. Where electronic circuitry is used for machine control it is equally important to keep it clean, dry, and free from dust. Many circuits depend for their functioning on tiny capacitors and resistors, soldered into the circuit board. The presence of dust or moisture can alter the values of these, causing failure or malfunction of the controller with its consequent effect on the moulding cycle.

Plasticizing unit

The plasticizing unit of the injection moulding machine consists of several components and includes: (1) A heating cylinder for softening and plasticizing the material and from which it can be injected into the mould; (2) an Archimedean screw which can both rotate and reciprocate within the cylinder; (3) a feed hopper; (4) a motor for rotating the screw; (5) a hydraulic cylinder and plunger for moving the screw forwards in the cylinder to effect injection; (6) a smaller hydraulic cylinder and plunger used for moving the carriage on which the plasticizing unit is mounted either forwards to bring the cylinder nozzle into contact with the mould, or backwards to provide clearance between these parts.

(1) *Heating cylinder.* Often referred to as the 'barrel', this consists of a robust steel cylinder capable of withstanding the pressure required for injection. In some machines, capable of limited usage, pressures up to

10 000 lb/in^2 (700 kg/cm^2) are used, but in the majority of machines, particularly those designed to mould some of the newer thermoplastics, up to three times this pressure may be developed. When it is considered that pressures of this order are similar to those developed during firing in the barrel of a naval gun, the need for robust construction becomes apparent. The cylinder is generally nitrided or may be lined with a hard alloy lining. The bore is accurately ground and honed to the required dimensions.

The cylinder is mounted on the headstock of the machine which, in turn, is mounted on a carriage, which can move forwards and backwards on lubricated slides. At the front end of the cylinder is fitted the nozzle—in effect, a reducing section which brings down the bore of the cylinder to that of the hole which communicates with the mould cavities. The various designs of nozzle are dealt with in Chapter 5 on mould design.

To assist in the removal of the screw from the cylinder, should the need arise, or to inspect the cylinder bore, some machines are designed so that the headstock can be rotated on its carriage in a horizontal plane, allowing the cylinder to swing free of the mould.

The cylinder is generally heated by means of electrical resistance heaters in the form of 'cuffs' which fit snugly around the barrel. The heat generated is conducted through the walls of the cylinder to soften the material inside. Small machines may have just one heater but larger machines have up to ten or more. Obviously, the larger the cylinder, the more heaters will be required, and the larger will be the wattage of the individual heaters. Ideally, the heating effect of each heater is individually controlled, but because of the expense in having large numbers of controllers, the heating is divided into zones and the temperature of each zone is thermostatically controlled. Thus, the heating zone at the rear of the cylinder, near the headstock is normally kept at a temperature below that at which material is completely plasticized, but serves to preheat it. In the centre of the cylinder the temperature is that at which the material is fully plasticized, and the front zone is controlled to keep the material at the correct temperature for injection. The nozzle is generally heated separately. In each zone there may be several heater bands and the placing of the control thermocouple and the choice of heater-band wattage become critical in order that uniform heating may be ensured. The drawback of zoning is that uniform heating is not easy to achieve and hot spots are liable to occur. A well designed cylinder heating system should have the minimum of hot spots and it should be possible to maintain a uniform temperature gradient along its length.

As already indicated, the temperature is normally maintained a little lower at the feed end than at the discharge end of the cylinder. There are exceptions. For example, if a material—soft PVC is one—relies for some of its plasticization on frictional heating, it may be desirable to have the cylinder somewhat hotter at the rear to reduce the friction and then, after the initial high temperature, to have a somewhat lower temperature until the material reaches the nozzle. It is then given a final boost before it enters the mould.

The placing of the control thermocouples and the characteristics of temperature controllers are dealt with later, but at this stage, one observation should be made: the functioning of an injection moulding machine is a cyclic operation. The screw which rotates inside the barrel also reciprocates and the areas where frictional heat develops also move backwards and forwards. Although one would expect this cyclic movement to even out any hot or cold spots there is always the possibility that the point in the cylinder where greatest frictional heating occurs may coincide with the point of greatest electrical heating. Because the controller only monitors the temperature of the cylinder wall and there is a layer of thermally insulating plastic between it and the screw, it is possible for the screw itself to become overheated. This is one of the causes of what is called 'screw slip' and the part of the screw affected does not pump (i.e. convey material) as effectively as it should.

(2) *Screw*. Like the cylinder in which it rotates, the screw is made of nitrided steel and is an accurate fit to the cylinder with, perhaps, 0·002–0·004 in (0·05–0·1 mm) clearance. The screw thread is generally of constant pitch (this is cheaper to machine than variable pitch) but the depth of the thread reduces along its length. (Or the diameter of the screw root increases from the back to the front of the screw.) The volume of a flight (one complete revolution of the screw thread) is therefore greater at the feed end than at the discharge end, giving compression and so compensating for the fact that the bulk density of the granular plastic is lower than that of the molten material. The ratio of compression is of the order $2\frac{1}{2}$ to 4:1 and the length/diameter ratio between 10:1 and 20:1, the most common ratio being about 15:1.

At its rear end, the screw is supported on a substantial thrust bearing and is fitted with a gear wheel in train with gears on an electric or hydraulic motor for applying rotation.

One of the most important features of the screw is the non-return valve on the front end which prevents material from flowing back along the flights when the screw is being used as a plunger. Two main types of non-return valve are used (Fig. 2.6a and 2.6b). Figure 2.6a is a check ring, which resembles a piston ring in an internal combustion engine, but seats on a chamfered section of the screw. When material is moving forward it pushes the ring off its seating and passes through gaps in a narrowed section of the screw root but, on injection, the pressure of the material pushes it firmly on to its seating. Figure 2.6b has a close-fitting plunger section with a minimal gap to the cylinder wall, fitted to the end of the screw. In this are holes in the rear of which are chamfered sections on which suitably sized ball bearings will seat. A serrated insert prevents the ball from being pushed forwards out of the hole.

The design of the screw thread is important because it determines the efficiency of the screw pumping action, the amount of mixing of the material in the screw flights and thus the degree of homogenization of the melt. A screw with a compression ratio of about $2\frac{1}{2}$:1, having a constantly decreasing depth and constant pitch, is the most suitable for materials which change from solid to fluid over a wide temperature range, such as polystyrene. Low density polyethylene, although having a crystalline melting point at about 112°C, also changes its melt viscosity gradually over a wide temperature range and is satisfactorily plasticized with such a screw. Materials such as nylons and acrylics, which either have a very sharp, high melting point and become very fluid, or pass through a phase of relatively high viscosity before becoming very fluid, are best plasticized with a screw having a very short compression zone. A screw of this type has a feed zone at the rear end of about ten flights of constant depth. At the forward end is a metering zone of about five flights of constant depth (but shallower than the feed zone). Compression takes place in between in the space of one half to one flight. The depth of the feed zone is of the order of $\frac{1}{6}$ of the screw diameter and in the metering zone about $\frac{1}{15}$ to $\frac{1}{25}$. There is a danger that, if a material having the characteristics of nylon were to be plasticized in a screw having a long compression zone, the flights could become blocked with solid material 'cemented' with semi-fluid plastic and feeding would not take place. There is also the possibility of breaking the screw in these conditions.

Different machinery manufacturers favour their own screw designs, most of which are produced with the object of providing a universal screw for use with all materials. Because possibly 40 per cent of all moulding is now repetitive work—that is, a machine is bought to produce one type of moulding in one specified material—rather than custom moulding, which requires maximum machine versatility, it would be profitable for the manufacturers to look again at machine design. For example, if the manufacturer concentrated on producing basic machines capable of doing one type of job in one material, with the option of adding to or modifying the basic machine to give more versatility if required, this might appeal to some purchasers. One feature to start with, when rationalizing design, is the screw. It is the component which is most difficult to make of universal application. It implies, of course, that screw removal would need to be very easy, accomplished without the need for external equipment, and capable of being performed by one engineer of no more than average skill.

(3) *Special screws and vented barrels.* Of recent years, vented barrel injection moulding has been much discussed and advertised. It would therefore seem opportune to consider this development before moving on to discuss the next element of machine construction. If the primary function of the screw and barrel is to plasticize material and convey it to the mould, its secondary function is also important. At the temperature of moulding, any moisture or other volatile matter in the moulding material is vaporized and must escape. To some extent, this vapour, with any trapped air, will be forced back along the screw flights and will escape at the feed pocket, but beyond a certain limit of volatile content, not all of it will be disposed of in this way. The volatile matter is carried forward with the molten plastic and is eventually released into the mould. There it condenses, forming the well-known splash marks or mica marks, described in Chapter 1. (Incidentally, by keeping the rear zone of the cylinder at a temperature which will allow only partial

softening of the granules, it is possible to mould slightly more moist materials because the more open structure of the material in the rear of the cylinder allows volatile matter to escape by this route somewhat more readily.)

By arranging for a screw to have a decompression zone about half to two-thirds of its length from the feed end, followed by a second compression zone, it is claimed by the manufacturers of the machines that quite moist materials can be satisfactorily moulded. A vent in the cylinder wall allows volatile matter to escape from the decompression zone before the plastic is re-compressed and passes into the mould. Claims are made that materials such as ABS, nylons, acrylics, polycarbonate, etc., can be moulded without predrying when using machines with vented barrels.

Great care is necessary in the control of temperature and feed because complete filling of all the flights, especially in the decompression zone, makes venting impossible. However, given care, the need for expensive drying equipment can be obviated. Whether or not really moist re-work (e.g., nylon type 66 sprues and runners which have reached equilibrium moisture content) can be effectively processed is not yet clear, and whether the additional cost of the longer barrel and more complicated screw can be justified is a matter for calculation. However, if the vented barrel proves, in the long run, to be all that is has been advertised to be, it will be a most significant development.

(4) *Feed hopper.* Not much needs to be said about the feed hopper except that it needs to be large enough! By that is meant that it should be of sufficient capacity to ensure that the operator is not continually having to refill it. (Most moulders have seen machines in which a bag of material has been opened and inverted into a feed hopper to give that component some extra and temporary capacity.)

The hopper should have a well-fitting lid to exclude moisture and dirt (some large moulding shops have an indigenous bird population!) and *the lid should be used.* A good sight-glass should be provided so that the material level can easily be checked and the hopper should have, overall, a smooth inner surface to which dust particles will not adhere. A sliding shutter at the base for cutting off the feed should be fitted and there should be some easy means of tipping or removing the hopper in order to empty it. There must be no constrictions in the feed throat across which particles could lodge to form a bridge and so stop the feed. Ideally, a magnetic screen should be fitted in the feed throat to trap any magnetic ferrous metal particles, but the hopper should be made demountable for cleaning and to enable the cylinder feed pocket to be cleaned. This is most important when changing colour or material, because the slightest particle remaining in the cylinder can contaminate a large number of mouldings.

Safety devices are not usually fitted to feed hoppers, but a screeen or grid to prevent anything—including a finger—being pushed through and getting trapped in the screw flights is desirable. A safety switch to prevent the screw rotating when the feed hopper is removed is also an advantage where personnel having some partial physical disability are employed.

(5) *Screw motor.* Two methods of providing rotation for the screw are used: an electric motor operating through a fixed or variable gear train, or a hydraulic motor working off the machine's hydraulic supply. It is most desirable to be able to vary the speed of the screw, as will be seen later, and this is most easily done by the hydraulic motor. The great disadvantage of the electric motor is the cost of fitting a fully variable-speed control. If a gear box with a number of gear ratios is used, it is not easy to change speed without stopping the motor and because the motor is of fixed speed, the torque applied to the screw is greatest at the lowest screw speed.

Hydraulic motors, on the other hand, give constant torque at all speeds and the speed can be infinitely varied by means of a flow control valve in the hydraulic supply line while the motor is running. It is desirable, although not essential, to have a screw speed indicator. This ensures that the optimum screw speed can be found and noted for any particular moulding job. There is no need for a very elaborate flow control valve—a simple needle valve will do—so long as the screw speed indicator is fitted.

The screw motor must have adequate power to rotate the screw against the friction developed between the semi-solid plastic, the cylinder walls, and the surface of the screw flights. One advantage of the hydraulic motor is that it will stall if the frictional resistance is too great whereas the electric motor will overheat or, if the gearing is low enough, could break the screw. A screw of 1 in diameter (25 mm) requires a 4 hp (3 kW) motor. One of $1\frac{1}{2}$ in (37·5 mm) requires 8 hp (6 kW) and the $4\frac{3}{4}$ in (135 mm) diameter screw for a nominal 100 oz (3000 g) machine uses a motor of nearly 100 hp (75 kW). For the most part, using hydraulic motors the screw speed can be varied from 0–200 r.p.m., but unless very rapid plasticization is required it is rarely necessary to use the highest speed. It is best practice to adjust the screw speed to the lowest necessary to retract and prepare the next charge in the time available. Too rapid a rate of rotation can cause frictional overheating of the material, the screw root becomes overheated and the effectiveness of the screw pumping action is impaired. Too slow a speed of rotation not only slows the plasticization rate, but also reduces the mixing effect in the screw flights, resulting in a lack of homogeneity in the melt injected into the mould.

(6) *Hydraulic injection cylinder.* Mounted on the headstock at the rear of the machine is a hydraulic cylinder and plunger, which is used for driving the screw forwards to bring about injection. The diameter of this hydraulic cylinder is much greater than the diameter of the screw so that it acts as a pressure intensifier. The pressure required for satisfactory injection is between 10 000 and 30 000 lb/in^2 (700 and 2100 kg/cm^2) and obviously it would not be feasible to utilize hydraulic line pressures of this order directly to bring about injection. A balance—generally greatly influenced by economics—has to be made between the desired hydraulic line pressure and the dimensions of the hydraulic cylinder required to achieve the necessary intensification ratio. For example, supposing a maximum injection pressure of 20 000 lb/in^2 (1400 kg/cm^2) is required on a 1 in (25 mm) diameter screw, the hydraulic

cylinder needs to be $3\frac{1}{4}$ in (84 mm) in diameter and the hydraulic line pressure 2000 lb/in^2 (140 kg/cm^2). With a line pressure of 1000 lb/in^2 (70 kg/cm^2) the cylinder diameter would need to be about $4\frac{1}{2}$ in (112 mm). In these examples the intensification ratios are 10:1 and 20:1 respectively.

It will be seen to follow from the foregoing that the volume of hydraulic fluid required to bring about injection is quite considerable. Thus, for a volume injection of 100 cm^3 (about $3\frac{1}{2}$ oz) using a line pressure of 1000 lb/in^2 (70 kg/cm^2) and an intensification ratio of 20:1 the volume of oil to be pumped is 2000 cm^3, about $\frac{4}{9}$ gal. Supposing that injection takes place in 1 s the flow of oil is 2 l/s or about 27 gal/min. If this volume of oil has to be directly pumped, a motor of about 25 hp is needed and it is for this reason that hydraulic accumulators in the hydraulic power circuit are almost essential.

Although, for the most part, the rate of injection needs to be as rapid as possible, and even the 1 s mentioned above may be 10 or 20 times too long for some mouldings, there are occasions when lower rates are desirable. Examples of jobs where slow injection may be needed are thick-sectioned mouldings in nylon and awkward shapes which are liable to give flow patterns on an appearance surface. In these cases, a flow control valve in the supply line to the injection hydraulic cylinder is necessary. The valve should be of good quality and compensated for temperature changes; its flow rate should be independent of temperature. It should also be capable of being set and reset accurately. It is very often found that, whereas the valve controlling screw rotation is quite sophisticated, that controlling injection speed is of much poorer design and is often located in an almost inaccessible situation on the machine. It could be argued that the flow control valve to the injection cylinder is one of the most important of machine controls and, as such, should be made more accessible because the ability to control the injection rate accurately can be advantageous where difficulties in mould filling are encountered.

(7) *Carriage hydraulic cylinder.* Often situated below the plasticizing cylinder, the hydraulic cylinder for moving the plasticizing unit on its carriage is an important item in determining the dry cycling time for the machine. To give adequate clearance between the mould and nozzle, for example, when changing the nozzle or cleaning a blocked sprue bush, the cylinder needs a fairly long stroke. The total force required for maintaining contact between nozzle and mould during injection depends on the size of the nozzle and the injection pressure. Typical figures are, for a 1 oz (30 g) machine, $1\frac{1}{2}$ tonnes and for a 4 oz machine, about 5 tonnes. The hydraulic oil pressure on the cylinder can generally be varied, as can the rate of flow, thus controlling the speed of movement of the carriage. For obvious reasons, the pressure should be only what is necessary to seal the nozzle/sprue bush joint and the speed of movement should be as rapid as possible, whilst avoiding damage to the parts coming into contact. These control valves should always be checked when setting the machine so that two common faults in technique are eliminated: nozzle leakage and excessive delay if sprue break is employed.

It is now obligatory in the United Kingdom to have adequate safety guards at all places on the machine where it is possible to harm the operator. There is a great risk of burning the hands if the carriage moves forward while the operator is working on the nozzle. On newer machines a guard is now being fitted which prevents access to the nozzle, except when the supply of oil to the hydraulic cylinder is switched off.

Mould unit

There are many types of mould unit—the part of the machine which the Hyatt Brothers called the 'hydraulic engine'. The main object of this part of the machine is to carry the mould sections, to open and close the mould, and to maintain it closed, without leakage of material during injection. A number of components are common to all designs and these will be dealt with in this section although mould design is considered in detail in Chapter 5.

The parts to be considered are (1) mould platens; (2) tie bars; (3) mould actuating and locking mechanism; (4) ejector mechanisms and (5) safety guards.

(1) *Mould platens.* Approximately in the centre of the machine is located the fixed platen of the machine. Cast in steel and of robust construction, it is firmly bolted to the frame of the machine in such a position that the cylinder nozzle will come into contact with it before reaching the end of its stroke. The centre point of the fixed platen should coincide with the centre of the nozzle, and there is normally a hole in the platen which varies in size according to the size of the machine. This hole is accurately machined to size on the side farthest from the nozzle and serves to locate a plate in the mould which exactly centres the mould to the nozzle. Typical diameters for the holes are 4 in (100 mm) for a 1–4 oz (30–100 g) machine and $6\frac{1}{4}$ in (160 mm) for a 40 oz (1200 g) machine. The hole is generally tapered outwards on the injection cylinder side to allow room for the front section of the cylinder to move close in to the platen when a very short nozzle is used.

At the rear of the machine is another robustly constructed fixed platen which is either freely resting on the machine frame or bolted to it with slotted bolt holes. This platen is held in position, relative to the other fixed platen, by means of the tie bars (see below). There are generally four tie bars, one in each corner of the platens. Intermediately between the two fixed platens is a third one which is free to slide on the tie bars, the holes in the platen having bushes which act as bearings. (The bushes, incidentally may be plain gunmetal or phosphor bronze; they may be impregnated with lubricant or have self-lubricating bearing material incorporated in them.) Like the other platens, the moving platen is robustly constructed and has a hole of 4 or 5 in (100–125 mm) in diameter in the centre. This allows for the installation of an ejector rod.

The assembly of the mould platens has to be made very accurately and the holes through which the tie bars pass must be absolutely vertical to the plane

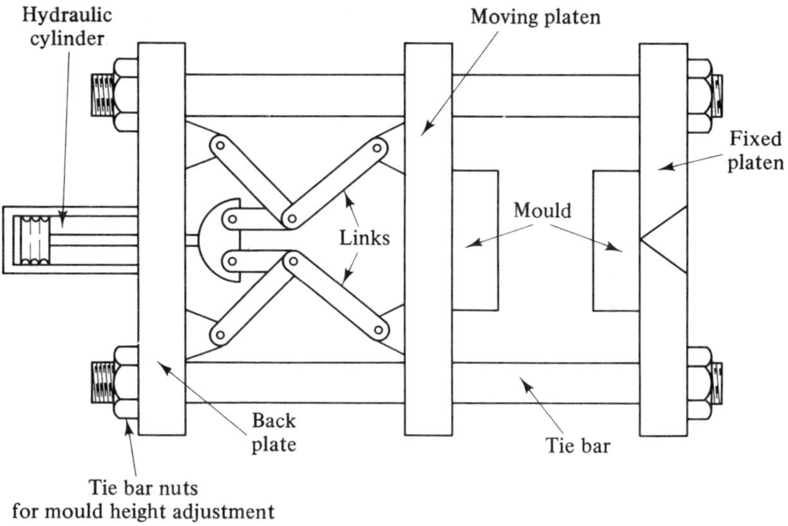

Fig. 3.5 Simple toggle lock

of the platens. The surfaces of the latter are ground perfectly flat so that the mould will fit without any distortion. The platens are not generally of solid construction—this would make them far too heavy—but are cast with appropriately placed cavities and buttresses so that they will present non-distorting surfaces. For mould mounting purposes, both fixed and moving platens are drilled and tapped to a European Standard pattern, or are furnished with undercut slots which will take the nuts of standard high-tensile bolts. The latter allows a little more versatility for mould mounting.

(2) *Tie bars.* Connecting the fixed platens, the tie bars, normally four in number, are made from high-tensile nickel/chrome steel. The surface is accurately ground and polished and there is a threaded portion at each end. Although there are exceptions, normal practice is for the tie bars to be firmly fixed at the fixed platen and for the back platen (sometimes called the back plate) to be able to move in respect of the centre platen by adjusting the position of the locking nuts on the elongated threaded portion of the tie bar. This adjustment is to allow for mould height adjustment (q.v.)

Typical stresses in the tie bars are of the order of 6–8 tonnes/in^2 (1000 kg/cm^2) and at this high loading, elongation can be as much as 1 per cent. This elongation is of importance in the design and functioning of the locking mechanism, as will be seen. It is allowed for in the machinery manufacturers' instructions for setting up the mould and it is unwise to deviate from them. The elasticity of the steel at the stresses encountered in injection moulding assumes a major importance.

(3) *Mould actuating and locking mechanism.* Between the moving platen and the back plate is the mould actuating and locking mechanism. There are

three main types in common use: (a) Toggle locking, (b) hydraulic locking and (c) a combination of (a) and (b).

(a) *Toggle locking.* In its simplest form, toggle locking is illustrated in Fig. 3.5. There may be a double toggle, as shown, or there may be single or quadruple mechanisms. Links are pivoted respectively on the fixed and moving platens and the two links are pivoted together with a third one. This is pivoted at its other end to a plate attached to the plunger of a hydraulic cylinder. As the plunger moves forward the link to which it is pivoted causes the other two links to straighten, this in turn pushing the moving platen away. The mechanical advantage of a toggle system can be as much as 50:1 so it is possible to obtain a very high locking force with a comparatively small hydraulic cylinder. In fact, in terms of hydraulic pressure it gives an intensification ratio of 50:1.

If the back plate were fixed in relation to the centre fixed platen it would be possible to use only one thickness of mould. So that moulds of different thickness may be accommodated (mould height adjustment) the back plate can be moved relative to the centre platen, hence the reason for the elongated threaded portion of the tie bars at the rear. However, setting the mould height becomes very tedious if all four tie bars have to be separately adjusted and it is almost impossible to obtain equal loading without the use of a strain gauge. Simultaneous adjustment of all the tie bars is brought about by linking all the nuts together, with either a heavy-duty chain and sprocket system or a gear train. This can then be turned manually or by means of an electric motor, and if the tie bars have been correctly set during manufacture there should be equal tension on all the tie bars.

This method, with modifications, is most commonly used and is probably the least expensive. A somewhat more sophisticated and, certainly more expensive method, involves fitting a fourth platen which is bolted to the machine frame and the back plate then becomes 'floating'. It is held relative to the fourth plate by a large, coarse-threaded screw to which it can be fixed and which passes through a tapped hole in the fourth plate. Mould height adjustment is made by rotating the screw and if the machine has been properly set during manufacture it is possible to obtain very even loading of the tie bars (Fig. 3.6).

Toggle locking mechanisms have several advantages. They give what is very desirable, namely, rapid intermediate movement of the platen with much slower movement at the beginning and end of the stroke. In moving large masses of metal there is always a good deal of inertia to overcome in starting and kinetic energy to absorb on stopping. The toggle mechanism prevents this energy from being absorbed in the mould, where it would cause damage, and from imparting wearing jolts to the machine generally.

In setting a toggle lock the link assembly is first backed off to allow the links to be fully extended. The mould is then hung on the fixed platen and the link assembly, with the moving platen brought up to the moving part of the mould, which is then bolted to the platen. The linkage is then collapsed and the back

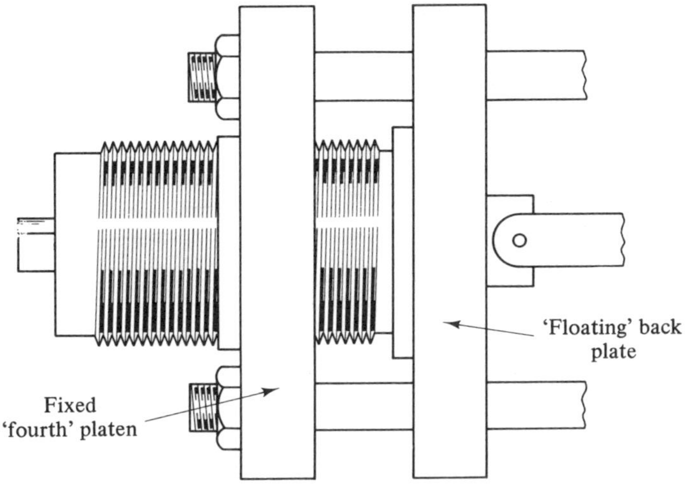

Fig. 3.6 Mould height adjustment for toggle lock

plate moved forward slightly. When the linkage is then extended again, full lock will be obtained if the central pivot can be taken very slightly past the fully extended point. The steel extends in an elastic manner to give the required lock. If the pivot goes too far over the fully extended point and a moulding is then made it will be found impossible to open the mould afterwards because the mould will have been forced open by the pressure of injection, the space between the mould plates will have filled with plastic, and the mechanism will not be able to compress the material enough to allow the links to extend fully before they fold and allow the mould to open. The only safe way of opening the mould then is to slacken off the back plate.

The main disadvantage of toggle lock machines is the fact that the pivot pins and bushes, being subjected to very high compression loading, tend to wear more quickly than can be tolerated. Obviously, a badly worn toggle mechanism will not allow good, uniform quality mouldings to be made, and replacement of worn bearings is both costly and time consuming. If great attention to correct lubrication is given with a high standard of cleanliness it is possible to obtain good performance for a long time. Careful setting and the use of no more locking force than is absolutely necessary to get good mouldings will also help to prolong the life of the bearings.

(b) *Hydraulic locking.* Hydraulic locking is the answer to some of the problems associated with toggle locking but it tends to have problems of its own. Figure 3.7 shows a typical arrangement and it will be seen that the back plate has been replaced by a large hydraulic cylinder attached to the frame of the machine. The plunger which operates in the cylinder is attached to the moving platen of the machine.

Mould opening and closing is generally effected by means of a much smaller cylinder and plunger, referred to as a 'jack ram', which, being small, uses much

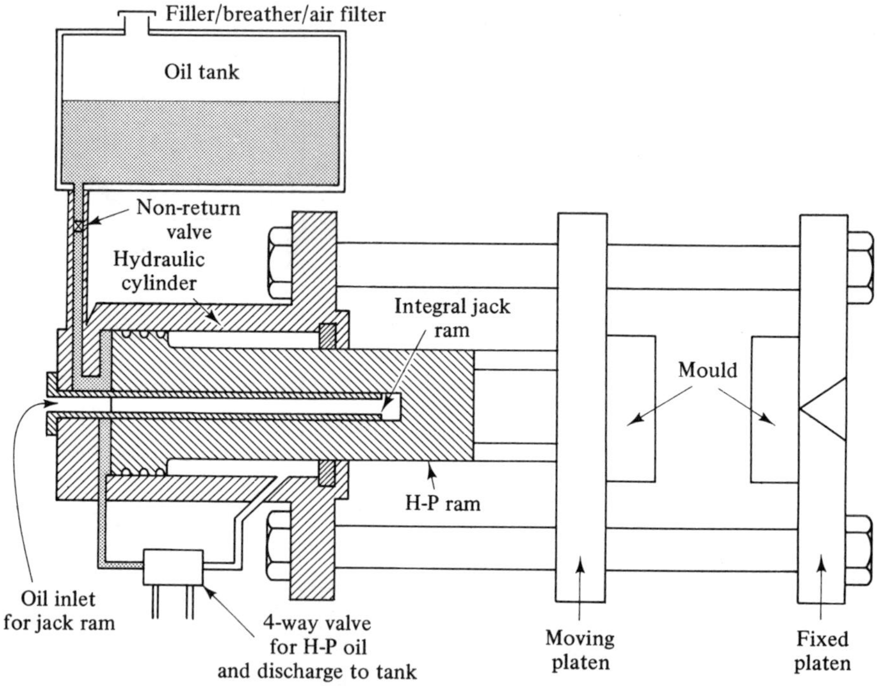

Fig. 3.7 Hydraulic locking using jack ram for opening and closing

less hydraulic oil and also gives a very modest locking force. The jack ram may be integral with the large plunger or it may be separate. In fact, on some large machines there may be two or more jack rams. After the jack ram has closed the mould, a valve operates and allows high pressure oil to enter the large cylinder which is already full of oil at atmospheric pressure. With the bigger hydraulic machines, the oil tank is very conveniently placed above the machine so that oil can enter the cylinder by gravity. A non-return valve in the line prevents high-pressure oil from flowing back into the tank when the pressure is applied. Only a relatively small quantity of high-pressure oil is then required for locking.

It is most important when using this system to make sure that all the parts of the mould can slide freely on their guide pins, that the platen moves freely on the tie bars, and that there is no resistance to movement. The small ram can then close the mould, the low pressure developed causes no damage to the mould if a moulding has stuck in it, and there will be no possibility of 'hobbing' the mould because of the one part being driven against another.

One problem that occurs with hydraulic lock machines is that, when resistance to mould closing occurs, the jack ram is unable to perform the operation. The mould safety system is then bypassed by the operator and mould closing is allowed to take place using the large locking cylinder. This, of course, is a rather hazardous procedure and can quickly cause serious trouble

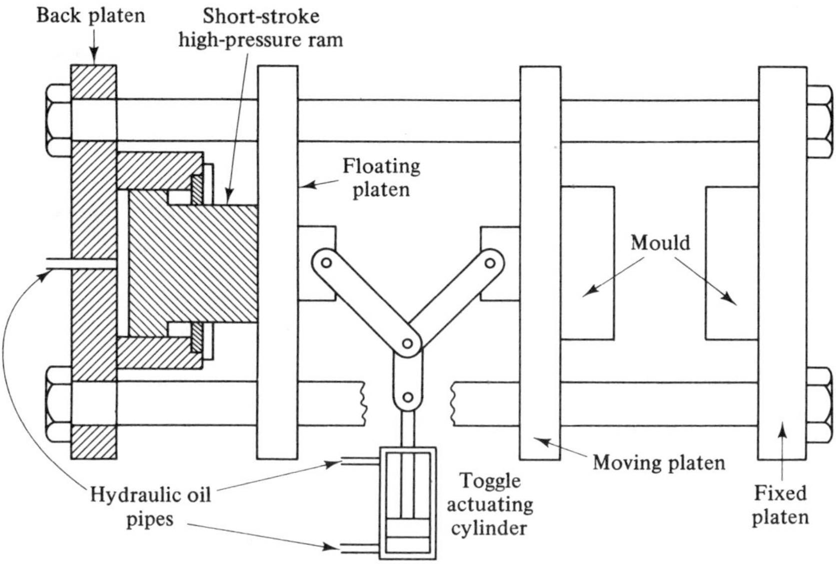

Fig. 3.8 Hydraulic locking with toggle opening and closing

to both machine and mould if allowed to continue. In this event, the mould should be re-set to ensure that it opens and closes easily.

Within fairly wide limits, mould height adjustment does not need to be made because the large stroke of the cylinder will accommodate it, but there is obviously a minimum limit for mould thickness. A stop is provided to prevent the hydraulic plunger being pushed too far along the cylinder. There is, however, one type of cylinder which does require additional mould height adjustment. This has a fairly short parallel section of cylinder at the forward end in which the plunger fits accurately to give locking pressure. Further back, the cylinder widens out so that the plunger moves freely without generating pressure. This arrangement gives valuable mould protection because the plunger/platen assembly can be moved forwards and backwards only by means of the jack ram.

Hydraulic lock machines tend to be slower in action than toggle machines but they have other advantages. Setting is easier because whatever position in the cylinder the pressure plunger takes up when the mould is closed (except with the cylinder just described), it is possible to lock the mould. Mould travel for ejection is easily adjusted—a saving of time when thin sections are being moulded—and tie bar stretch is accommodated by the fully adjustable locking system. Variations of tie bar length due to temperature fluctuations are automatically accommodated.

(c) *Combined toggle and hydraulic locking*. The advantages of both systems with few of the disadvantages of either can be obtained by combining toggle opening and closing with hydraulic locking. Figure 3.8 illustrates the system. Mould height adjustment can be made as for a toggle machine or by the fitting of standard spacers. It does not need to be more accurate than say, $\frac{1}{4}$ in (6 mm) because the final lock is taken by a large-diameter small-stroke hydraulic cylinder and piston. Mould movement is rapid, as for a toggle machine, with slowing at each end of the stroke. This system overcomes the problems of tie bar stretch and of variations in tie bar length due to temperature changes. Mould safety is assured because virtually no pressure is applied until the mould is almost completely closed (the toggle-actuating cylinder can be very small). There is little movement of hydraulic fluid and cycling can be very fast. Until about 1974 only very few machinery manufacturers in the world were using this system (it may be somewhat more expensive than the others) but by 1978 more suppliers are incorporating hydraulic pressure pads in toggle machines to give final locking.

(4) *Ejector systems.* Types of moulding ejectors will be discussed in Chapter 5 on mould design, but mention should be made here of the two systems, mechanical and hydraulic ejection, which are an integral part of the mould system. Air ejectors are a separate issue and are considered under mould design.

Provision is usually made for either a central rod or a bar which passes through the mould and operates the mould ejectors to be moved by some means attached to the mould actuating mechanism. For central ejection, a threaded rod screwed into a fixed member in the centre of the machine comes into contact with the central ejector rod on the mould as the mould opens. The length of the rod can be adjusted to give the movement required. For use with an ejector bar in the mould, there are usually two adjustable rods, one on either side, so that as the mould opens equal contact takes place on either end of the ejector bar.

Mechanical ejectors of this type are cheap to instal but they can cause a great deal of unnecessary wear and tear on the machine. They are often adjusted to come into action at a time when the platen movement is very fast. This causes jolting of the mould and the machine and the rapid movement of ejector pins or the like against the moulding can cause tearing or distortion.

Hydraulic ejection can help to overcome this problem. A hydraulic cylinder with a short-stroking plunger is mounted centrally in the moving platen. The plunger movement causes it to contact the ejector bar or rod in the mould and so to eject the mouldings. The hydraulic oil supply to the ejection cylinder can be switched either mechanically with microswitches on the moving platen or by time switches integrated into the machine sequencing controls. It is therefore possible to bring about ejection of the mouldings at any time after the mould starts to open; the stroke of the cylinder can be adjusted to suit the ejector movement required, and the speed of ejection can be regulated to suit the article being moulded. Jolting of the mould and machine, which occurs with mechanical ejectors, is eliminated and there is considerably less wear and tear and less damage to mouldings. It is also possible to arrange for multiple-stroke ejection (sometimes called 'shaking').

(5) *Safety guards.* The mould end of the machine is the most dangerous from the point of view of accidents to the operator and to personnel setting up the machine. Adequate guarding is essential. Access to moving parts must be prevented during operation and this applies to the mould, the actuating mechanism, and to the chute under the mould down which the finished mouldings fall. Double safety devices, both electrical and mechanical, are very desirable and should interlock so that the mould cannot be closed whilst the guard is open. Some machines will allow for the guard to be opened before the mould opens at the end of a cycle, and although this may be satisfactory with a fully competent operator, there are advantages from the point of view of safety in making it impossible. It could be possible to trap a finger or hand between the moving platen and the guard or other part of the machine framework which, although not so dangerous as being trapped between the mould parts, could be a painful experience.

A mechanical device such as a stout steel bar which drops down to prevent the mould parts from closing when the guard is open is an added safety precaution. If extensive work is to be performed on a mould *in situ* in the machine, it is always wisest to stop the pump motor and to make sure that any accumulators are isolated. It is also good practice when working on one side of a mould to protect the other side by hanging a piece of board or strong plastic sheet over its face.

Not only is personal safety of importance, but provision should also be made to give the mould added protection. The most accurate and careful setting of the mould safety devices is essential, but even so there is always a possibility that a moulding will lodge somewhere in the outlet chute and foul the mould. Photoelectric and mechanical devices for checking that a moulding has, in fact, been ejected can give almost 100 per cent certainty, but there is still no really effective substitute for the timely action of a vigilant and well-trained operator.

Machine specifications

Consideration should now be given to the machine specifications and the claims that machinery manufacturers make for their products. These specifications are conveniently divided into three parts: plasticizing end, mould end, and power supply.

While there is no point in using a larger machine than is necessary to make a particular moulding, neither is there any virtue, apart from cost, of always trying to match machine size to moulding. There may be overriding reasons for the choice of machine, such as the amount of mould opening which is necessary to eject a small-weight moulding if it is, for example, in the form of a long tube. A small thick-walled article may not require a large lock but a large machine may be required because of the shot weight. Similarly, it may seem perfectly feasible to make a certain moulding on a particular machine, but if the machine is not capable of performing the injection in the time which can be allowed, the quality of the mouldings—if they can be made at all—will almost certainly be inferior.

Manufacturers' specifications always present the best view of the machine and it is therefore wise to check some of the claims from first principles. Obviously, there is not a deliberate attempt to mislead, but figures are notorious for being manipulated to tell the desired story. A manufacturer's typical published specification is as follows:

Plasticizing end	
Screw diameter	40 mm
Maximum shot weight (polystyrene)	81 g
Swept volume (80 per cent efficient)	76 cm^3
Injection pressure (maximum)	1960 kg/cm^2
Injection time	1 s
Maximum screw drive power	6 kW
Maximum screw speed	195 r.p.m.
Stroke of screw	76 mm
Heating zones on barrel	3
Heating zones on nozzle	1
Total heating capacity	6·25 kW
Hopper capacity	22 kg

This is a fairly full specification but there are some omissions. For example, the injection time of 1 s does not indicate a *rate* and whether or not the pressure remains constant. It is necessary to calculate this from the hydraulic motor capacity. Other information which is lacking is the wattage of the individual heating units in each zone and for what material 22 kg is the rated hopper capacity. It is also useful to have an indication of the average screw-back time for a specific material. Information on the plasticizing capacity which is often given is not very useful, as this can depend on so many other factors such as screw design and speed and the amount of frictional heat developed.

Mould end	
Clamping force	80 tonnes
Overall size of platens	480 × 457 mm
Space between tie bars	305 × 270 mm
Mould height adjustment	150–300 mm
Opening stroke (adjustable)	50–228 mm
Tie bar diameter	60 mm

Other details which could be of use are the minimum times required for opening and closing the mould. These are of value when estimating the overall time cycle for any project because they are the only parts of the time cycle not capable of being calculated.

The last part of the specification may be very full and informative or, like the one being quoted, may give just one piece of information:

Power unit	
Main drive motor	15 kW

This information is, of course, essential but other items of importance include:

Capacity of oil reservoir
Total installed power requirements
Hydraulic line pressure (maximum)
Hydraulic line pressure (minimum or working)
Cooling water requirements

These items are all relevant to machine choice, running costs, installation costs, etc.

Also normally given, often with an outline sketch of the machine, are the overall dimensions and weight. This is very useful when deciding on a position for the machine and determining the floor strength or lifting tackle capacity which will be required. It is important, too, to know whether a machine has a built-in hydraulic accumulator. Often, this is not mentioned or even implied and the enquirer is forced to conclude that no accumulator is fitted.

Machine control

Mention has already been made of switches, relays, and solenoids. Also necessary for the functioning of the machine are process timers and temperature controllers.

(1) *Process timers.* Two main types are used in injection moulding, electric/clockwork mechanisms with mechanical switches, and solid-state timers. There are two main variations of the former type: one can be adjusted to give timing intervals of, say, 6, 60 or 600 s simply by altering a scale and gear train, and the other operates over one particular interval such as 0–10, 0–30 or 0–100 s. Solid-state timers have no moving parts and usually have a 3-digit indicator which can be set at any time from 0 to 99·9 s. At the end of the time set either type of timer can be made to operate a relay which, in turn, will actuate a solenoid.

Mechanical timers generally have a pointer to indicate the passage of time and this is an advantage because it is an indication of how far through the particular part of the cycle the machine has gone. Digital timers only indicate the start and finish of the time period by means of indicator lights. Both types are extremely reliable, but digital timers are slightly easier to set and are nearly always more legible.

If it is necessary to know the elapsed time during a cycle of operations an elapsed-time digital indicator can be fitted. At least one make of machine has this attachment as standard equipment and it can be set to indicate not only the whole cycle but the various parts such as injection time or mould opening time. These times can be of importance in fast cycling operations where the loss of fractions of a second can make a difference to the profitability of the process.

One of the most desirable attributes of a timer is that it should be readily visible and easy to read. There seems to be little virtue in putting timers into a panel situated where only a stooping dwarf can read them. They should be visible and legible to the operator standing by the mould guard.

(2) *Temperature controllers.* Three main types of temperature controller are used on injection moulding machines: (a) Simple energy regulators, (b) on–off indicating controllers, and (c) proportioning and energy regulating controllers.

(a) Simple energy regulators. The first type includes rheostats, variable-voltage transformers and on–off time proportional switches. These controllers are chiefly used for nozzles, hot-runner moulds, etc., where it is desirable to control the heat input, but where there is no need to have a temperature indicator once running conditions have been established. They are also used for the heating cylinders of some small or bench model machines, where more expensive control systems cannot be justified, in which case a simple thermometer may be used to indicate the temperature.

(b) On–off indicating controllers. On–off indicating controllers are in use in many machines, more particularly older models. A thermocouple, set in a pocket in the cylinder wall, produces a small current which operates a sensitive millivoltmeter. The instrument is suitably calibrated and there is a movable pointer which is adjusted to the set point temperature required. This pointer carries a photoelectric cell and a small lamp. When the temperature of the cylinder reaches the set point the pointer of the voltmeter moves to indicate it and a flag which it carries comes between the lamp and the photoelectric cell, cutting off the current it develops. This operates a relay which switches off the current to the heater bands. When the temperature falls, the photoelectric cell is re-activated and the current is again switched on.

With an on–off controller the position of the thermocouple is of importance. If it is deep-set in a pocket in the cylinder wall it will indicate the temperature of the melt to which it is close, but because the heat takes a long time to penetrate the thick metal the temperature fluctuates or ‘hunts’. Set in a shallow pocket near to the heater band the controller switches on and off frequently and the heat is supplied to the machine in small ‘doses’. The melt temperature stays much steadier but can be several degrees lower than that indicated on the instrument.

(c) Proportioning controllers. Proportioning controllers which modulate the signal current so that when it is rising or falling the switching comes earlier also reduce ‘hunting’. The thermocouple can then be in a half-way position, where it gives a better indication of melt temperature. Energy regulation can be added by means of thyristor pulse switching, so that the heating current is varied according to requirements. When the temperature is far from the set point full current flows, but as it approaches the set point it gradually reduces to a level which will just maintain the temperature required. This type of instrument is more expensive than the on–off controller, but having no moving parts it is extremely reliable and maintains a very steady temperature.

Process sequencing

The operation of an injection moulding machine commences with the switching on of the main power supply, starting the flow of cooling water, and provision of any auxiliary services. The heater circuits are generally separate from the motor supply, which is nearly always 440 V, three-phase. Heaters are 220 V, single-phase, as are other electrical supplies to mould heaters, etc. Moulding should not be attempted until all temperatures have reached those required for normal running, and during the warm-up period heater controllers should be set to the correct level.

Most machines are equipped for fully-automatic, semi-automatic, and manual operation. It is always advisable to take off a shot at 'manual' before switching over to 'automatic'. Assuming the machine has been left in the proper shut-down position, the mould will be open and the screw will be in the forward position. The feed hopper and screw flights will be empty. First, the operator should make sure that the nozzle is clear. In starting the heating of the cylinder it is wise to have the nozzle heated first, so that if any decomposition takes place in the cylinder the products, which may be gaseous and liquid, can escape through it and not through the cylinder and feed hopper. The feed hopper can now be filled and the cylinder charged.

The carriage is moved back so that there is clearance between the nozzle and the sprue bush on the mould. Screw rotation is started and continued until plasticized material issues from the nozzle. The screw rotation is stopped and the screw is brought forward, as in injection, to purge the first fluid plastic from the cylinder. This operation should be repeated until any dirt is eliminated. The purgings should be collected on a tray for waste disposal. When the purgings appear to be clean, moulding can start.

Pressure and flow control valves are set to the correct positions for mould operation and injection, and it is assumed that all the safety switches, sequencing switches, and feed controls have been properly set. The mould guard is then closed. This actuates the mould closing mechanism and when the mould is closed the carriage is moved forward to bring the nozzle into contact with the sprue bush. The injection button is then depressed and the screw moves forward to cause the mould to fill; after a few seconds, screw rotation is started and the screw rotates until its backward motion operates the feed control switch. The mould can then be opened and a moulding ejected. Any modifications to pressures and flow control valves are made and the machine is switched over to semi- or full-automatic.

Any adjustments to temperatures, pressures, and times should be made gradually and moulding quality carefully checked. It is better to start with a somewhat longer cycle than necessary, especially on fast-cycling jobs, and gradually to reduce the times, rather than hope to achieve optimum working conditions straight away.

A typical full order of sequencing is as follows:

(1) Signal to close mould.
(2) Low pressure mould closing.
(3) Signal to lock mould.
(4) Mould locks.
(5) Signal to move carriage forward.
(6) Carriage forward, nozzle in contact with mould.
(7) Signal to inject. Timer starts for injection.
(8) Injection takes place until pressure reducing valve operates[1] then continues under reduced pressure.
(9) Timer signals end of injection, screw starts rotating, and cooling timer starts.
(10) Screw rotates until cam operates feed limit switch to stop rotation.
(11) Signal to retract screw further to give decompression of melt (optional).
(12) Signal to withdraw carriage to predetermined limit (optional).[2]
(13) Cooling timer signals mould to open.
(14) Mould opens and operates ejector.
(15) Mould fully open, pause timer starts.
(16) Pause timer signals mould to close and operation repeats.

It has been said that when injection moulding is running smoothly it is the production engineer's dream of happiness. Unless close attention is paid to the detail as outlined in this chapter it is very likely to turn into a plastics technician's nightmare!

[1] This can be done either with a timer or a cam and microswitch.

[2] Carriage withdrawal (sprue break) should be minimal and should take place only if the nozzle tends to cool off or if there is difficulty with sprue removal.

4

Materials

Much has been written of the chemical nature and origins of plastics and it is not intended to repeat here more than is necessary for the understanding of the behaviour of the materials which are processed by injection moulding. It is not possible to injection-mould all plastics. Some, such as cast acrylic and PTFE (polytetrafluoroethylene), cannot be made to flow freely enough to allow of processing by this method. Others, such as mixtures of resin and glass fibre in woven or mat form are unsuited by their physical nature for use in the process. In general, a plastic, whether of the *thermosetting* or the *thermoplastic* type, which is capable of being brought to a state of fluidity similar to that of molasses or treacle, can be injection moulded. What this means in rheological terms will be discussed later in the chapter.

Thermosetting and thermoplastic materials

The division of plastics into the two main types, thermosetting and thermoplastic, is both traditional and scientifically useful. Some of the first commercially produced moulding materials, based on phenol and formaldehyde, were thermosetting. On heating, such materials become fluid and can be shaped under pressure in a mould. As heating proceeds, a chemical change, catalysed by suitable additives, causes the moulded material to harden and after a suitable time it can be removed from the mould without distortion, even though it is still hot. In chemical terms, what happens is that the molecules of which the material is composed, although originally separate entities, become chemically bonded together. The reaction is called cross-linking and imparts to the moulded piece its insoluble and infusible nature. Such cross-linked material cannot be re-ground and used again.

Thermoplastic materials, on the other hand, always remain capable of being softened by heat and of hardening on cooling, even after repeated cycling. The molecules of which they consist always remain as separate entities; they do not cross-link to any great extent although during the moulding process they undergo interesting physical and rheological changes.

Molecular or fine structure of plastics

Nearly all the plastics at present commercially used are based on compounds of carbon and hydrogen. Oxygen, nitrogen, chlorine, fluorine and silicon may also be present in some plastics. The general behaviour of the whole range of materials can be explained by reference to one of them—polyethylene or polythene—which is, chemically speaking, the simplest of the thermoplastics.

The element carbon (chemical symbol C) is found in nature uncombined in three forms: soot or lampblack, an amorphous form—it has no crystal structure; graphite, a crystalline form in which the crystals are plate-like; and diamond, the cubic crystalline form. Combined with other elements such as hydrogen, oxygen and nitrogen, carbon is found in all living tissue. Hence the name 'organic' for the compounds of carbon. In petroleum, carbon occurs combined with hydrogen as the hydrocarbons, of which methane, or natural gas, is the simplest. In this compound, one atom of carbon is united with four atoms of hydrogen. Carbon atoms can also join together in chains or rings, but always the combining power (or valency) of carbon, which is the equivalent of four hydrogen atoms, must be satisfied. Methane is given the formula CH_4 (H being the chemical symbol for hydrogen); other well-known hydrocarbons have formulae as follows:

Propane (an industrial gas)	C_3H_8
Butane ('Calor' gas)	C_4H_{10}
Octane (in petrol)	C_8H_{18}

A hydrocarbon oil such as is used for lubricating might have a formula $C_{18}H_{38}$ and it will be seen that, as the number of carbon atoms in the formula increases, so the nature of the material it represents is different, from the gaseous methane to the oily, waxy C_{18} compound. The chain of carbon atoms may not always be 'straight'; it may be 'branched' thus:

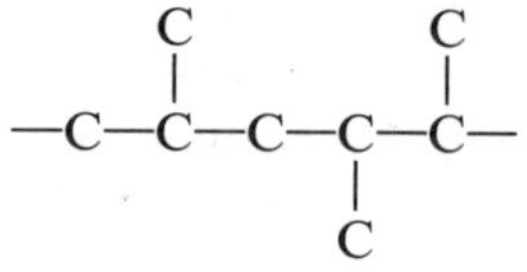

and with octane there are 18 possible arrangements of the atoms. All have the same basic composition but individually they all have slightly different properties. For example, they might have slightly different boiling points or their chemical activity may be slightly different.

Polyethylene, or polythene, has many more carbon atoms in its 'skeleton'. A simple grade such as is commonly used in injection moulding may have 2000 carbon atoms linked together and combined with 4002 hydrogen atoms. If a molecule of this type took up the form of a straight chain it would have a length/diameter ratio of 2000:1 and its total length would be 12/1 000 000 in (3/10 000 mm). X-ray and electron microscope studies have shown that polythene molecules lie side-by-side some 4/10 000 000 mm apart and that, although the term 'straight chain' is used, the actual atoms of carbon are arranged in a zig-zag formation as shown below in which the angle θ is 112°:

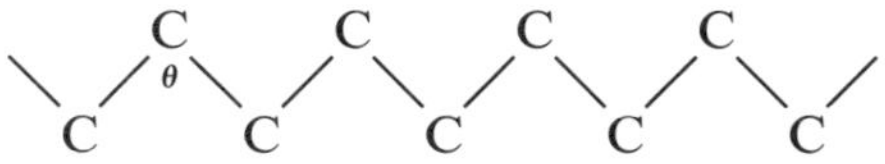

This description of the molecular structure of polythene may seem to be of only academic interest, but it is the arrangement and disposition of the molecular chains that determine the behaviour of the material during moulding and explain many of its properties. Four aspects of this behaviour should be noted.

(1) Having such a large 'aspect ratio' (length/diameter), it will be readily appreciated that entanglement of molecules can take place. The larger the molecule and the more branched the chain, the greater will be the entanglement and thus the greater the forces needed to make one molecule slide against another. Greater forces will be required to rupture or to shear the material. Thus, as the molecular weight (the relative weight of all the atoms in the molecule compared with that of one atom of hydrogen) increases, so the strength increases and the melt viscosity (the resistance to flow of the melt) also increases. Note that the 'melt flow index' which is a measure of the rate at which the material flows through a die, *decreases* as molecular weight increases.

(2) The molecules of polythene form a regular structure and, so long as there are not too many short side chains, about 50 per cent of the material will settle into the most convenient regular pattern and form into crystals. These are quite small, about 3/100 000 mm in diameter, and are of spherulitic shape. If the chains have less branching the possibilities of crystallization increase and with some types of polythene well over 90 per cent can be in this form. Because the crystal structure is more closely packed than the non-crystalline material it is denser, giving the so-called high-density polythene. This type of polythene has a density of 0·96 g/cm^3 compared with 0·92 g/cm^3 for the 50 per cent crystalline material. Some of the other effects of crystallinity will be discussed later.

(3) With the molecules having the zig-zag structure mentioned above, it could be supposed that, given enough stress, the chain could be drawn out by almost 20 per cent in length, that is, the bond angle could be increased from 112° to 180°. This would require immense forces and no doubt slipping of one molecule past another would occur before the 'straightening' neared its limit. In fact, this probably happens to the material in the molten state when first viscous flow takes place under shear and finally the molecules are drawn out in an elastic manner. Under shear in the mould some straightening or 'orientation' takes place and if the molecules cannot relax before solidification occurs an extension of about 2 per cent may be frozen in. After removal from the mould the material will still try to relax—the molecular forces involved are very great—and, unless restrained by a heavy rim section, distortion of the

component will occur. If there has been some measure of differential extension, such as is normal in a centre-fed moulding, the typical saddle back distortion which is very common in cheap household mouldings, will take place.

(4) The normal distance between the individual molecules is fixed by a balance of forces. There is the force that holds the molecules together to give the solid mass and there are also repellant forces arising from their chemical nature tending to push them apart. By applying a strong external force it is possible to make the molecules lie more closely together and this is seen when the molten plastic is extruded rapidly through a small orifice. This is the kind of situation found in the gate to the mould and its effects will be considered at the end of this chapter.

So far we have considered the arrangement of the molecules in a simple plastic, consisting of carbon and hydrogen only, arranged in substantially linear manner. There are more complex plastics which consist of the same elements arranged in a somewhat different way and there are other plastics in which elements such as chlorine or fluorine are joined to the carbon atoms. Still other plastics consist of groups of elements linked together in a regular fashion by 'linking groups' consisting of small units containing carbon and oxygen, carbon and nitrogen, or silicon and oxygen. For a more detailed study the reader is referred to a book such as 'Plastic Materials' by J. A. Brydson published by The Butterworth Group.

Chemical structure of plastics

Direct substitution of one of the hydrogen atoms on alternate carbon atoms in the basic polyethylene molecule by a group of atoms consisting of one carbon atom and three hydrogen atoms gives the well-known plastic, polypropylene. In this material there is an added complication in that the substituting groups may be arranged either on the right-hand or left-hand side of the main chain. There are three possibilities in a plastic of this kind and they are known as stereo-specific polymers. The three arrangements are:

(a) Random, in which there is no fixed pattern of arrangement.
(b) Alternate left-hand and right-hand substitution.
(c) Regular, in which the substituting groups are always on the same side.

It so happens that (a) type polypropylene (it is given the description 'atactic') is a mastic and is no good on its own as a plastic. (b) is a good plastic but is difficult to make economically and is called 'syndiotactic'. (c) is easier to make and is a good plastic, the description being 'isotactic'. Because it has a regular structure, isotactic polypropylene can be up to 95 per cent crystalline but because of the side groups the molecular chains will not pack so closely as those of polythene and it has a lower relative density (0·9). Polypropylene bears some resemblance to polyethylene but is much stronger and stiffer than the low-density variety.

The other important plastic which consists entirely of carbon and hydrogen is polystyrene. This can be thought of as a basic polythene-type chain molecule having on alternate carbon atoms a group of carbon and hydrogen atoms arranged in a six-membered ring. Although this side group is bulky, it is also relatively dense and the polystyrene plastic has a density of 1·06 g/cm^3. The substitution is generally in a random manner and polystyrene does not form into a crystalline structure.

By substituting chlorine atoms for hydrogen on alternate carbon atoms the well-known plastic polyvinyl chloride (PVC) is obtained. Chlorine atoms are heavy compared with hydrogen and the resultant plastic has a density of about 1·4 g/cm^3 and because the chlorine atoms are also much more bulky, the chain molecules cannot move so freely and the strength and rigidity of polyvinyl chloride are much greater than those of polythene. The presence of chlorine atoms in the molecule also increases the resistance of the plastic to burning. On the other hand, heating PVC to a temperature only slightly above its processing temperature tends to cause decomposition. Hydrochloric acid is liberated and this tends to accelerate the rate of decomposition.

The nylon group of plastics, of which there are five or six types in common use, are typified by having short hydrocarbon chains linked by groups consisting of carbon, nitrogen, hydrogen, and oxygen atoms. This type of group is hydrophilic, i.e. it tends to absorb water, and its presence in the nylons causes the moisture absorption which is well-known in those materials. The hydrocarbon groups do not attract moisture, so the proportion of linking groups in the various types of nylon determines the moisture absorption. This is greatest in nylon types 6 and 66 and much lower in nylon types 11 and 12. Nylon type 66 is a somewhat anomolous material. It has strength and rigidity in excess of what would normally be expected and its melt viscosity is very low. These characteristics can be explained by the fact that it is possible, because of its structural peculiarities, for a loose kind of chemical bonding to occur between adjacent molecules, causing them to pack more closely (the density is about 1·14 g/cm^3) and for less molecular entanglement to take place.

Acetals contain oxygen atoms in addition to carbon and hydrogen, but the oxygen acts as a link between adjacent carbon atoms in the chain. It does not cause the material to absorb moisture but it does make it less resistant to heat—in fact it supplies some of the oxygen which is necessary for oxidation in the event of burning—and the nature of the chemical bonds allows it to break down fairly easily into the simple chemical formaldehyde. This is a rather pungent smelling vapour which has a decided lachrymatory effect. The so-called acetal copolymers have the alternate carbon–oxygen chain interrupted by an occasional carbon–carbon linkage and this tends to make the materials less sensitive to heat and slightly more stable.

Other plastics—polycarbonates, polysulphones, polyphenylene oxide and thermoplastic polyesters—all have more complicated chemical structures and the aim in 'designing' these plastics is to derive a chemical type and structure that will give a set of desired properties. This is not always quite simple, but there is another expedient available to the plastics chemist which makes it possible to 'tailor make' a material for a particular application, at any rate, within certain limits. This is the process of copolymerization, and to understand this it is necessary first to consider the method by which plastics are made.

Polymerization and copolymerization

The normal method employed by the chemist for making a plastic is to take a fairly simple but reactive compound (in the case of polythene this is a hydrocarbon gas called ethylene) and to cause the small molecules to join together into the long chain structure. It is also possible to take simple chemical compounds which have two reactive groups and to cause them to join together into long chains, often with the liberation of small molecules such as water. These processes are called polymerization and the products are polymers. The simple starting materials are known as monomers. By varying the conditions of reaction—temperature and pressure—or by having more or less catalyst present, it is possible to vary the properties of the polymer. Thus, it can be made with a greater or lesser molecular weight, and the distribution of portions of polymer with higher or lower molecular weight can be varied as also can particle shape and size. Where it is possible to make stereo-specific polymers their types can be varied and it is also possible to vary the amount of chain branching that occurs. Even this degree of control may not be sufficient to enable the correct combination of properties to be obtained and copolymerization is resorted to.

In copolymerization, two or more monomers are polymerized together in the desired proportions. Thus, a small proportion of ethylene copolymerized with propylene gives a propylene copolymer which has slightly modified properties compared with the ordinary polypropylene or homopolymer. Similarly, copolymers of vinyl chloride or of styrene can be made, and sometimes two copolymers may be blended to give an even better product. Examples of copolymers are:

(a) EVA (ethylene/vinyl acetate), a rubbery plastic having some of the properties of polythene but modified to suit applications where greater flexibility is needed.

(b) SAN (styrene/acrylonitrile), a tough, transparent plastic combining the rigidity of polystyrene with the toughness of poly-acrylonitrile (this plastic is used in the textile industry for yarns of great strength and a 'feel' similar to that of wool).

(c) ABS (acrylonitrile/butadiene/styrene), a tough 'engineering-type' plastic made by blending copolymers of butadiene and acrylonitrile and butadiene and styrene. It has the rigidity of polystyrene, combined with the toughness and chemical resistance of acrylonitrile, modified to give high impact strength by the presence of butadiene (this latter is one of the main components of synthetic rubber).

All the materials described above are derived from petroleum or petroleum products, such as naphtha, and their manufacture is an important part of the petrochemical industry. Although differing in chemical make-up, their structures all bear a resemblance to that described for polythene and the general considerations with regard to processing them by injection moulding are similar. Differences of melting or softening temperature and of their thermal, thermodynamic, and rheological properties occur. The greatest differences in behaviour are, however, shown between two groups of plastics, those having a crystalline structure in the solid state and those with a non-crystalline, or amorphous, structure.

Crystalline and amorphous plastics

It is theoretically possible, as has already been implied, by the use of stereo-specific catalysts during polymerization, to make all plastics into a form which will crystallize. This is not always expedient. For example, PVC normally has a random structure and does not crystallize. It is reported that a regular-structured polymer has been produced, but that its melting point is above that at which PVC decomposes so, until a suitable stabilizer can be developed, a crystalline type of this plastic would not be of any use. However, many of the plastics in common use have naturally crystalline structures and can be processed at well above their melting points. It is useful to consider the main differences between these and the amorphous plastics, and Table 4.1 notes some of the differences.

Table 4.1 Properties of crystalline and amorphous plastics

Property	Crystalline	Amorphous
Melting or softening	Fairly sharp melting point	Softens over a range of temperature
Density (for the same material)	Increases as crystallinity increases	Lower than for crystalline material
Heat content	Greater	Lower
Volume change on heating	Greater	Lower
After-moulding shrinkage	Greater	Lower
Effect of orientation	Greater	Lower
Compressibility	Often greater	Sometimes lower

Typical crystalline plastics are:
Polythene, polypropylene, nylon, acetals and thermoplastic polyesters
Typical amorphous plastics are:
Polystyrene, acrylics, PVC, SAN and ABS

These differences in properties are reflected in the behaviour of the materials during injection moulding. Taking first the sharp melting point of the crystalline plastics, as typified by nylon 66 which melts at 265°C: below this temperature it is a rigid, solid material of high strength but at less than 10° above its melting point it is a fluid material having a viscosity of about 1000 poise (100 Ns/m^2), well below the minimum viscosity required for injection

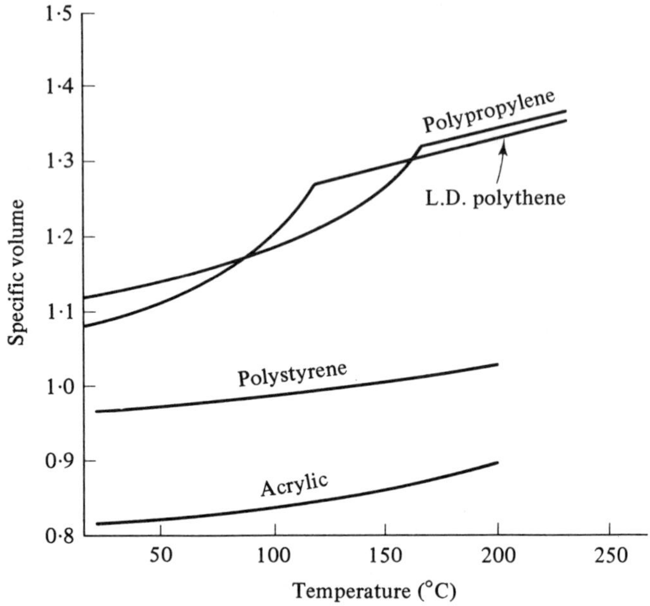

Fig. 4.1 Specific volume vs temperature

moulding. Similarly with polypropylene: at 167°C it melts to become a fluid material well within the range of viscosities required for injection moulding.

Perhaps the most important practical effect of this rapid change of state on heating and cooling is that mouldings can be removed from the mould at higher temperatures than those made from materials having no sharp melting point and a reduction in time cycle is possible. This will be discussed again later.

The second difference, that of density, provides a simple method of checking variation in mouldings made with crystalline materials. Any alteration in moulding conditions which affects the degree to which crystallization takes place in the mould—mould temperature, differential pressures, etc.—will make slight differences in the density of the material and this can be checked by simple physical methods.

When considering the effects on injection moulding of the third difference, the increased heat capacity of crystalline materials, two matters are of importance. First, the rated plasticizing capacity of the machine is often based on polystyrene, which has a low heat capacity, and this means that the machine may not be able to handle as much crystalline material as indicated by the given rating in a given time. The difference may be as much as 30 per cent. Second, the mould has to be designed to that the extra heat can be removed and cooling chanels may have to be larger or more numerous.

Another factor, in which safety is involved, is sometimes overlooked. When a material such as nylon melts it becomes very free flowing and it has been known

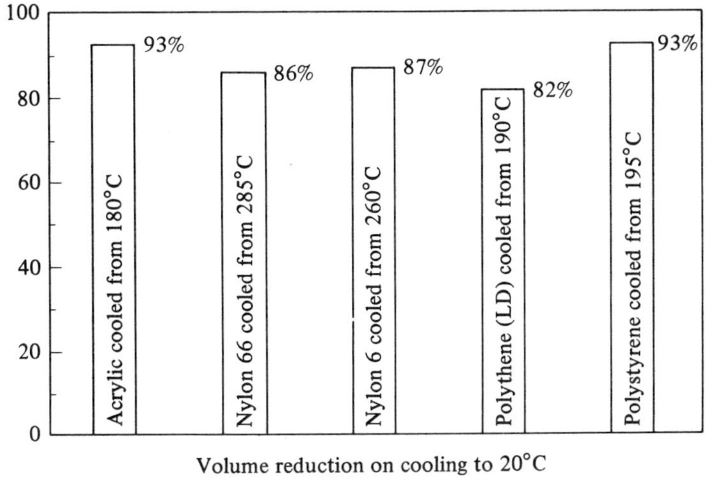

Fig. 4.2 Volume reduction on cooling

for it to spurt out from the parting surfaces of the mould. Adequate guards made from transparent sheet material are therefore recommended. A splash on the skin from a material of such high heat capacity as nylon can be extremely dangerous.

All plastics undergo a considerable volume change on heating, but the change with the crystalline ones is much greater than with amorphous materials. This is shown graphically for some typical materials in Fig. 4.1, where specific volume is plotted against temperature. It will be seen that whereas amorphous plastics change by an average of about 0·05 per cent in volume for every °C change in temperature, crystalline materials change by about 0·11 per cent.

For the practical moulder, the same information is presented in Fig. 4.2 and it will be seen that, in cooling the materials from their normal moulding temperatures to room temperature, volume changes of 7 per cent for amorphous materials and between 14 and 18 per cent for crystalline materials occur. This means that crystalline materials are more likely to give sink marks and voids than are amorphous ones, but it also indicates that a considerable volume of extra material at the moulding temperature needs to enter the mould in order to make up for the reduction in volume that takes place on cooling. How this is done is a matter of using the other properties of the materials, such as compressibility and behaviour on shearing.

Sometimes crystalline materials tend to show a fault in mouldings made from them known as windows. Normally, a natural coloured, crystalline plastic such as polythene, nylon, or polypropylene, will be translucent because the crystallites have a slightly different refractive index from that of the amorphous material and the light rays passing through are scattered. Amorphous material of these plastic types is generally clear and transparent, unless pigmented.

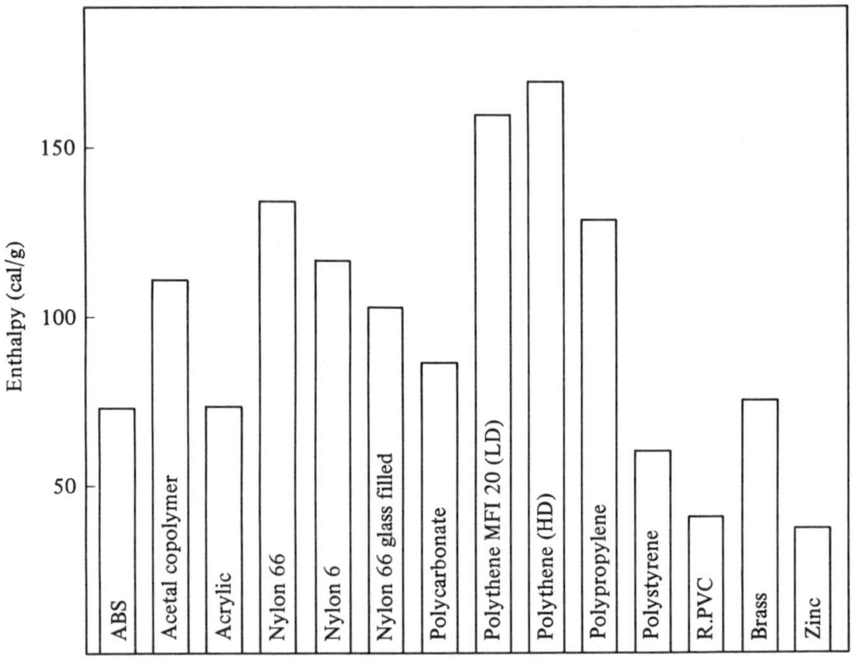

Fig. 4.3 Enthalpy of plastics and metals at moulding temperature (compared with 20°C)

The 'window' occurs because a granule of the plastic in passing through the cylinder takes up only enough heat barely to break down the crystal structure. It then enters the mould and is suddenly cooled before the molecules have time to re-form into the crystalline form and a piece of supercooled amorphous material is left in the moulding. Such a state of affairs only occurs when the plasticizing capacity of the cylinder is not great enough or the temperature too low and when insufficient time is allowed for thorough plasticization.

One other aspect of crystalline materials compared with amorphous ones is shown in their tendency to have greater after-moulding shrinkages. This is, to some extent, caused by the formation of the more closely packed crystal structure in the moulding, but this does not account for the fact that the shrinkage in the direction of orientation is about twice that at right-angles to it. Now, it could be considered that crystalline plastics having the more streamlined configuration will allow of greater extension of the polymer chain on shearing than will amorphous ones for any given shear stress. There is obviously more entanglement of amorphous molecules than of crystalline ones and the amount of polymer chain extension will therefore be less. Since it is the amount of chain extension that influences after-moulding shrinkage, crystalline plastics would be expected to shrink more than amorphous ones, particularly after undergoing high shear stresses.

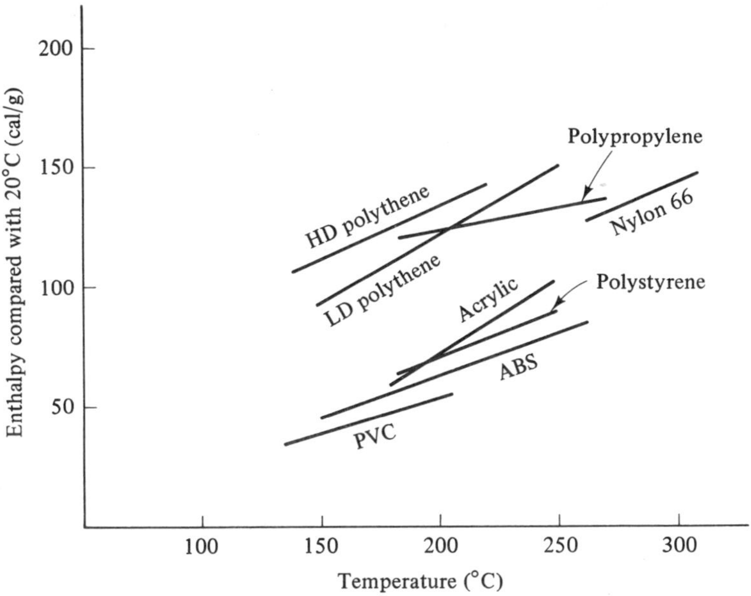

Fig. 4.4 Enthalpy of plastics vs temperature (compared with 20°C)

Thermal properties of plastics

Compared with metals, all plastics have a high heat capacity (specific heat). At their normal moulding temperatures, the total heat content of plastics compared with the heat content at, say, 20°C can be greater than that of zinc or brass at their melting points. The heat content, often referred to as the 'enthalpy', of some plastics at their moulding temperatures is shown in Fig. 4.3 and is compared with that of brass and zinc. This heat content can be 'put into' the plastic as well as being 'taken out' and the former process takes place in the cylinder, the latter in the mould. Figure 4.4 plots the enthalpy of some plastics against temperature and it will be seen that the crystalline materials such as polythene, polypropylene, acetals, etc., have heat contents at a given temperature which exceed by some 50 cal/g (209 J/g) those of the amorphous plastics, such as polystyrene, PVC, etc., at the same temperature.

In injection moulding, which is a highly capital-intensive process, the duration of the time cycle is of primary importance in costing. The dilemma is that the heat required for full plasticization is very great and the rate at which it can be put into the plastic is limited by the thermal conductivity and the thermal stability of the material. When melting most metals it is feasible to apply a very hot flame to the solid, knowing that the heat will be rapidly conducted throughout the mass. With plastics, which have very low thermal conductivities, this is not possible, because the high temperature on the outside would cause burning and decomposition before enough heat could penetrate to melt the mass of material. Table 4.2 gives some figures for the

Table 4.2 Thermal conductivity of plastics and metals (cal/sec cm °C)

ABS (acrylonitrile/butadiene/styrene)	6×10^{-4}	Polypropylene	$3{\cdot}3 \times 10^{-4}$
		Polystyrene	$2{\cdot}0 \times 10^{-4}$
Acetyl (homo- and copolymer)	$5{\cdot}5 \times 10^{-4}$	Polyvinyl chloride	
		rigid	$4{\cdot}0 \times 10^{-4}$
Acrylic	$4{\cdot}5 \times 10^{-4}$	flexible	$3\text{–}4 \times 10^{-4}$
Nylon 66	$6{\cdot}0 \times 10^{-4}$	Cellulose acetate	$4{\cdot}8 \times 10^{-4}$
Nylon 610	$5{\cdot}3 \times 10^{-4}$	CAB	$4{\cdot}7 \times 10^{-4}$
Nylon 11	$8{\cdot}8 \times 10^{-4}$	Copper	1·043
Nylon 66 glass filled	$7{\cdot}0 \times 10^{-4}$	Brass	0·204
Polycarbonate	$5{\cdot}3 \times 10^{-4}$	Steel	0·115
Polythene LD	8×10^{-4}	Aluminium	0·502
Polythene HD	$11{\cdot}5 \times 10^{-4}$	Zinc	0·264
Polyphenylene oxide	$5{\cdot}2 \times 10^{-4}$	Wood (oak)	0·0006

thermal conductivities of plastics and other materials but the bar graph in Fig. 4.5 illustrates this more dramatically. From this graph it will be seen that, whereas it takes 1 s for a unit of heat to be conducted through 1 cm cube of copper with a temperature difference of 1°, it takes about 1 h (3600 s) for the same amount of heat to be conducted through a typical plastic under the same conditions.

It must also be remembered that about half the heat put into the plastic in the cylinder has to be removed from the moulding before it can be taken out of the mould. It therefore follows that the hotter the plastic becomes in the cylinder, the longer the moulding will need to stay in the mould and therefore the longer will be the overall time cycle.

An indication of the maximum temperature at which the cylinder heaters

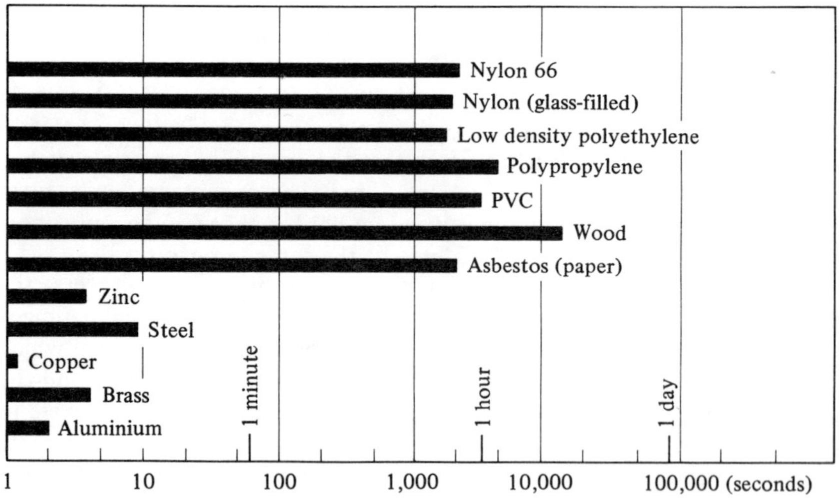

Fig. 4.5 Time to conduct 1 cal of heat through 1 cm³ of material with temperature difference of 1°C

Table 4.3 Processing and decomposition temperatures

Material	Processing temp. °C. (approx.)	Onset of decomposition °C. (approx.)
ABS	225	310
Acetal	200–220	220–240
Acrylic	200–225	250
Nylon 66	300	310–320
Polystyrene	200	250
Polythene	200–280	320–350
Polypropylene	220	280
PVC	150–210	180–220

can be set for any given plastic is given by the approximate temperature at which decomposition begins to assume undesirable levels. Table 4.3 gives such figures for a number of materials and also indicates some typical moulding and suggested cylinder temperatures. In interpreting these figures it should be borne in mind that the amount of decomposition which takes place is normally a function of both temperature *and* time. (Exceptions, perhaps, are acetal and PVC, where the decomposition can reach a point where it seems to become self-catalysing.) In injection moulding it is often necessary to compromise. For example, using a shot weight only slightly less than the maximum for the particular cylinder and with fast cycling, it might be possible to use a somewhat higher cylinder temperature than that indicated in Table 4.3. However, in the event of a machine stoppage it would be necessary to reduce the temperature settings quickly to prevent decomposition from taking place. Conversely, using a long moulding cycle and a smaller shot weight on the same machine, it would be necessary to run with a much lower cylinder temperature to reduce any possibility of decomposition.

Calculations with respect to heat transfer in plastics processing make use of the concept of thermal diffusivity. The largest proportion of heat flowing through a plastic material is by diffusion or conduction. Little heat is transferred by convection (the melt is too viscous) and radiation, although having some effect, accounts for only a minor proportion of the heat transfer. Mixing takes place as the screw rotates in the cylinder, and, in addition to electrical heating through the cylinder walls, there may be some generation of heat by mechanical means. Cooling in the mould is almost entirely by diffusion.

The coefficient of thermal diffusion is given by:

$$\alpha = k c_p^{-1} \rho^{-1} \ (\text{cm}^2\ \text{sec}^{-1})$$

where α is the coefficient of thermal diffusion (thermal diffusivity)
k is the thermal conductivity (cgs units)
c_p is the specific heat
ρ is the density (g/cm^3)

Some figures for thermal diffusivity are given in Table 4.4.

Table 4.4 Thermal diffusion constants ($cm^2\ sec^{-1}$)

Polythene LD	$0{\cdot}9 \times 10^{-3}$
Polypropylene	$0{\cdot}9 \times 10^{-3}$
Acrylic	$1{\cdot}0 \times 10^{-3}$
Polycarbonate	$1{\cdot}1 \times 10^{-3}$
Nylon	$1{\cdot}3 \times 10^{-3}$

(All at usual moulding temperatures)

In 1959, Ballman and Shusman (*Modern Plastics*, McGraw-Hill, New York, November 1959) gave an account of experimental work in determining moulding 'set-up' time (that is, the time for a moulding, in the mould, to become hard enough to remove) and put forward an equation by which it could be calculated:

$$Q = \frac{-D^2}{2\pi\alpha} \log_e \left\{ \frac{\pi}{4} \frac{T_x - T_m}{T_c - T_m} \right\}$$

where Q is the set-up time in seconds

D is the maximum thickness in centimetres

α is the thermal diffusivity

T_x is the heat distortion temperature

T_c is the cylinder temperature

T_m is the mould temperature.

Results obtained experimentally for polystyrene showed good agreement with calculated times using the formula and this gives a useful method of estimating cycle times for moulding. Of course, the mould configuration and the effectiveness of the mould cooling arrangements can have overriding effects, but provided that the latter have been well designed, the calculated results will be reasonably close to those obtained in practice.

In order to make the formula more universally useful, especially when considering crystalline plastics, a further amendment must be made. As indicated above, the figure for thermal diffusivity is calculated from a formula which includes a figure for specific heat. Normally, and this is acceptable for materials such as polystyrene, the value for specific heat would be that obtained by experiment at 20°C. To take into account the latent heat of fusion of crystalline plastics, the specific heat should be amended to the average value over the whole of the cooling range, taking the enthalpy at moulding temperature compared with that at 20°C, and dividing this by the temperature differences. Table 4.5, which gives the main thermal properties of a wide range of plastics, includes these amended values for thermal diffusivity and the figure can then be inserted directly into the formula for calculating set-up time. Some materials manufacturers are now able to supply data on the thermal diffusivity of their products.

The temperature at which a moulding can be removed from the mould

Table 4.5 Thermal properties of plastics

Material	Density g/cm^3	Specific heat	Latent heat of fusion cal/g	Moulding temperature °C.	Total heat cal/g	Thermal diffusivity cm^2 sec^{-1}/°C
ABS	1·01	0·35	—	225	72	$1{\cdot}7 \times 10^{-3}$
Acetal (homopolymer)	1·42	0·35	39	210	105	$0{\cdot}7 \times 10^{-3}$
Acetal (copolymer)	1·41	0·35	39	225	111	$0{\cdot}72 \times 10^{-3}$
Acrylic	1·18	0·35	—	225	72	$1{\cdot}09 \times 10^{-3}$
Nylon 66	1·14	0·40	31	280	135	$1{\cdot}01 \times 10^{-3}$
Nylon 610	1·13	0·40	26	245	116	$0{\cdot}90 \times 10^{-3}$
Nylon 11	1·05	0·58	40	220	156	$0{\cdot}95 \times 10^{-3}$
Nylon 6	1·13	0·38	38	250	125	$0{\cdot}98 \times 10^{-3}$
Nylon 66 GF	1·38	0·30	22	290	103	$1{\cdot}33 \times 10^{-3}$
Polycarbonate	1·20	0·30	—	300	84	$1{\cdot}47 \times 10^{-3}$
Polythene LD						
MFI2	0·92	0·55	50	280	193	$1{\cdot}17 \times 10^{-3}$
MFI 7	0·918	0·55	50	260	182	$1{\cdot}15 \times 10^{-3}$
MFI 20	0·916	0·55	50	220	160	$1{\cdot}09 \times 10^{-3}$
Polythene HD	0·94	0·55	50	240	171	$1{\cdot}57 \times 10^{-3}$
Polyphenylene oxide	1·06	0·32	—	310	93	$1{\cdot}53 \times 10^{-3}$
Polypropylene	0·90	0·46	24	250	130	$0{\cdot}65 \times 10^{-3}$
Polystyrene	1·05	0·32	—	200	57	$0{\cdot}6 \times 10^{-3}$
PVC (rigid)	1·44	0·24	—	180	38	$1{\cdot}16 \times 10^{-3}$
PVC (flexible)	1·30	0·40	—	160	56	$0{\cdot}7 \times 10^{-3}$
Cellulose acetate	1·28	0·36	—	195	63	$1{\cdot}04 \times 10^{-3}$
CAB	1·19	0·35	—	200	63	$1{\cdot}27 \times 10^{-3}$
SAN	1·09	0·33	—	220	66	$0{\cdot}81 \times 10^{-3}$

Notes: Moulding temperatures are average values
Total heat is compared with 20°C
Thermal diffusivity figures allow for latent heat of fusion

Table 4.6 Typical temperatures and factors

Material	Cylinder °C	Mould °C	Heat distortion[1] °C	Factor
ABS	230	50	80	190
Acetal	225	100	140	305
Nylon 66	280	70	180	140
Nylon 66 GF	300	130	230	81
Polythene	230	30	90	200
Polypropylene	250	60	120	338
Polystyrene	200	30	90	340

Moulding set-up time (seconds) = factor × max. thickness (cm)^2

[1] Typical maximum temperatures for ejection (i.e. average overall temperature of the moulding).

without danger of distortion is quite arbitrary. The published heat distortion temperature is a good guide, but with some materials and some types of component could be much too low, giving an excessively long set-up time. Table 4.6 gives some suggested temperatures which can be used in the formula for mould, cylinder and heat distortion and in the final column is a factor. This factor is a simplification of the whole of the equation with the exception of the

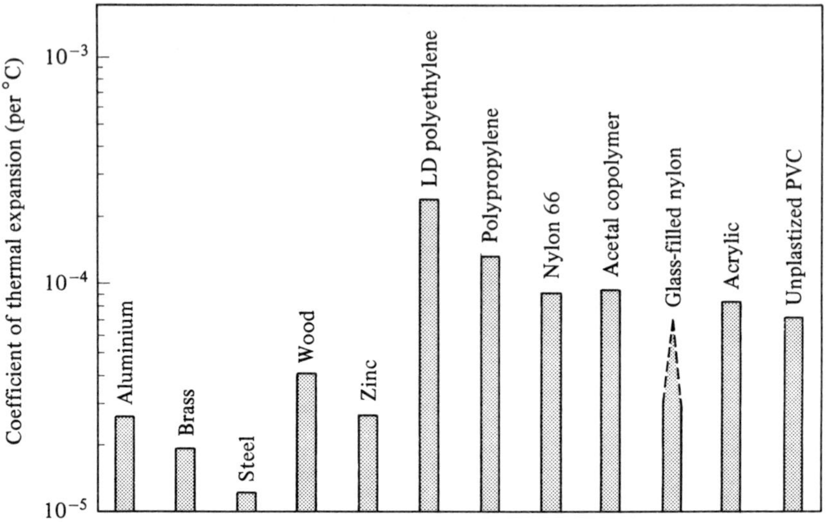

Fig. 4.6 Thermal expansion of metals, wood, and plastics

maximum component thickness, using the figures for the variables as quoted in Tables 4.5 and 4.6. By multiplying the factor in the last column by the maximum component thickness in centimetres, the set-up time can be calculated directly. However, with mouldings thicker than 0·5 cm or of varying thicknesses, there may be some disparity between calculated and observed times.

Taking the factor for polystyrene as 340 and assuming a component thicknesses of 1 mm, the set-up time is:

$$340 \times 0{\cdot}1 \times 0{\cdot}1 = 3{\cdot}4\ \text{s}$$

To obtain the overall cycle time, the injection time, the mould opening and closing times and the pause time must be added.

It is often observed that, when changing a colour, using the same type and grade of material, it is necessary to change the settings for the cylinder temperature. In general, whites, blues, and greens require a slightly higher setting than red, brown, purple, and black. This difference could be attributable to the different effects of radiation on material heating, as one would expect the black, brown, and red hues to absorb more radiation than the lighter colours. It can be assumed that the diffusion constant is the same for all colours. Any differences are small, but they can be significant and temperature settings may need to be adjusted by as much as 5–10°.

The final effect of heat on plastics which should be considered at this stage is the thermal expansion. Figure 4.6 gives some comparative figures for plastics, wood and metals. There are considerable differences in thermal expansion even between plastics, ranging from a coefficient for glass-filled nylon of about 2×10^{-5}/°C to over 2×10^{-4}/°C for polythene. A useful average figure for

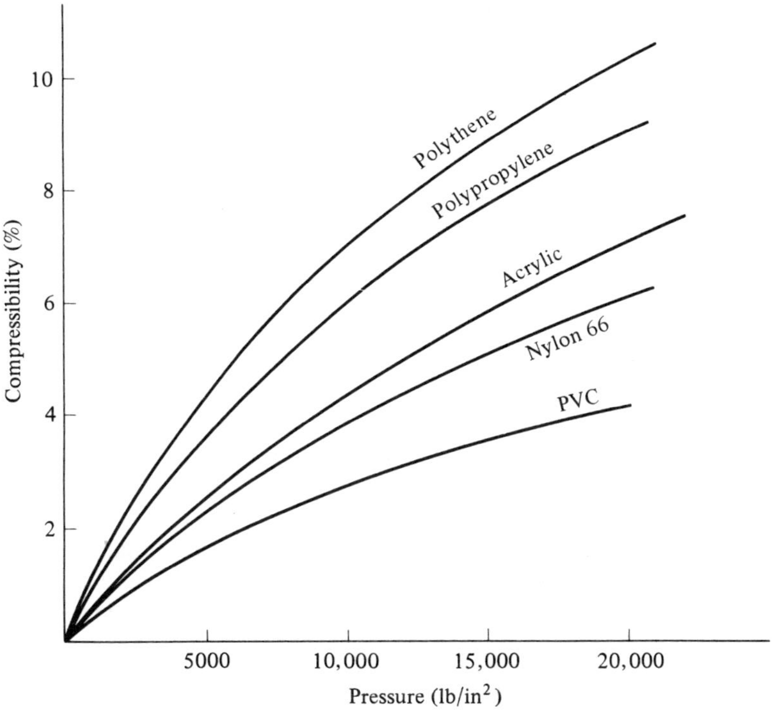

Fig. 4.7 Approximate compressibility of plastics

the British engineer to memorize is in terms of 'one thou per inch for every ten degrees centigrade or eighteen degrees fahrenheit). It is essential, when setting moulding tolerances, to bear these figures in mind because, although the moulder can justifiably boast to be able to produce mouldings to tolerances of $\pm$zero, this costs money. The overriding effect of thermal expansion on a component that can be expected in service may make tight tolerancing nonsensical.

Effect of pressure on molten plastics

It is often thought, erroneously, that by pressurizing the material in the mould it is possible to make up for the reduction in volume in the mould as the component cools. By doing so, it is hoped to avoid sink marks and voids in the moulding. From the figures already mentioned for the change in specific volume which takes place as the plastic is heated or cooled, it is seen that it would be necessary to compress amorphous plastics by over 7 per cent and crystalline plastics by double that amount to compensate for the reduction in volume as the moulding cools.

Figure 4.7 indicates the compressibility of five typical plastics at temperatures near to those required for moulding, and it will be seen that, even at

22 000 lb/in^2 (1500 kg/cm^2) polythene and nylon are not compressed by anything like what is required, acrylic falls short of its required 7 per cent and only polystyrene exceeds what is needed. Since the mould will almost certainly be pressurized to only about 10 000 lb/in^2 (700 kg/cm^2), because of pressure losses between the plunger and the mould cavity this method of producing sink-free mouldings would not not seem to hold out much hope of success.

Extreme pressurization of the mould, in general, leads only to localized areas of frozen-in stress in the moulding because it is only rarely that the pressure will even itself out in a completely filled mould. The viscosity of the melt is generally too great and the rate of cooling of the material too rapid to allow this to happen. It is best, in injection moulding, to use only the pressure necessary to enable injection to take place at the desired speed and for only the duration of injection. The pressure can then be held at only what is necessary to prevent back-flow from the mould until the gate freezes. After this, holding the plunger forward is completely unnecessary. However, if very thick-walled mouldings are being made, it is permissible to inject at a slow rate over a long period—in fact, flow moulding, in which the screw continues to rotate during injection, has met with some success—but there is always a danger of other faults developing. With materials such as acetal, continuing to 'feed' the mould to cut down sink marking and voids can cause splay marking.

Mouldability of molten plastics

Various attempts have been made from the early days of plastics processing to assess the 'mouldability' of the materials. The cup flow test for thermosetting moulding powders is one such attempt and the spreading disc test for thermoplastics is another. Among the more useful, though purely empirical tests, is the so-called method of spiral flow moulding.

Developed in the 1950's, this test involves moulding into an open-ended channel marked off in centimetres or inches, and arranged, for convenience, in the form of an Archimedean spiral, hence the name. Moulding pressure is maintained constant and, for a particular material or range of materials, moulded spirals are produced at various cylinder temperatures. (The test can also be carried out at constant temperature and the pressure varied, but the results obtained are less meaningful.) The results obtained are plotted on a graph to show spiral length against temperature (Fig. 4.8).

Spiral flow curves serve a useful purpose in indicating which materials require higher pressures, higher temperatures or accurate control of both temperature and pressure. A glance at the graph will show that acrylics require high pressures because the maximum spiral length is fairly small. Nylon 66 obviously requires a high temperature and, because the curve is so steep, accurate control is indicated. Polythene, however, is mouldable over a very wide range of temperature with only a comparatively small change in spiral length. The graph also indicates which materials can be used for purging—that is, scouring one material out of the cylinder with another. Reference to

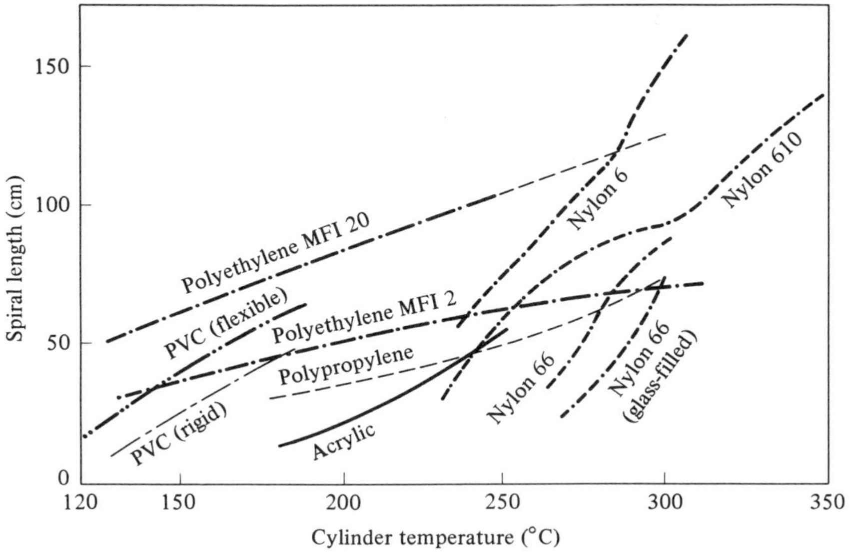

Fig. 4.8 Spiral flow curves

the graph shows that at 270°C, nylon 66, being 'stiffer' than polythene will purge it, but that at 300°C polyethylene will be stiffer than nylon 66 and can therefore be used to scour it from the cylinder.

In so far as making a quantitative assessment of the moulding characteristics is concerned, spiral flow moulding is not the complete answer. Mention has already been made of the way in which the molecular structure of the plastic can be distorted by high shearing forces, and this affects the behaviour of the material during injection into the mould. This behaviour can only be quantified by rheological methods.

Rheological properties of plastics

Rheology is the science of the flow of materials. It involves the motion, considered in terms of laminar flow, of one plane of material against another, and the shearing forces involved. The term 'viscosity' is used to define the force involved and is the tangential force per unit area acting on either of two horizontal planes at unit distance apart, one of which is fixed, while the other moves with unit velocity.

Consider a plate of area A moving against another plate with a velocity v at a distance apart of r. If a force F is needed to cause the movement then:

$$\text{Shear stress} = F/A$$
$$\text{Shear rate} = v/r$$

and viscosity, or resistance to flow, is:

$$\frac{\text{Shear stress}}{\text{Shear rate}} = \frac{F \times r}{A \times v}$$

Assume A is 1 cm, $v = 1$ cm/s and r is 1 cm:

$$\text{Shear stress} = F/1\ \text{cm}^2$$

$$\text{Shear rate} = \frac{1\ \text{cm}}{1\ \text{s} \times 1\ \text{cm}} = 1/\text{s}$$

$$\text{Viscosity} = \frac{F \times 1\ \text{s}}{1\ \text{cm}^2 \times 1}\ (F \text{ in dynes})$$

The unit of viscosity is the 'poise':

$$1\ \text{poise (P)} = 1\ \text{dyne s/cm}^2$$

It follows that the viscosity of a fluid can only be measured when it is being sheared. With some fluids, the viscosity remains constant at all rates of shear and the fluid is said to be Newtonian. Water, glycerine, and sugar solution are almost perfect Newtonian fluids (none is absolutely perfect), but melted plastics are not. Plastics fluid melts show a decrease in apparent viscosity as the shear rate increases and are said to exhibit pseudoplastic properties. Materials which show the same type of behaviour are the so-called jelly paints and the type of margarine which 'spreads straight from the fridge'.

In order to characterize the flow behaviour of plastics melts it is necessary to state both the shear rate and the temperature, because the viscosity depends on both variables. Viscosity measurements are made in a high shear rate viscometer, which consists of a heated cylinder, a die of known dimensions in the base, and a plunger which can move in the cylinder. The plastic to be tested is placed in the cylinder, allowed to heat gently to a predetermined temperature and then pressure is applied by means of the plunger to cause the plastic to extrude through the die.

There are various methods of using this type of equipment, two of which are indicated below:

(1) The plastic is extruded through the die with a constant load on the plunger and the amount of material extruded in a given time is weighed or estimated. The amount in grams extruded in 10 min through a standard die at a standard temperature is known as the melt flow index (MFI) of the material. It certainly gives some indication of the viscosity of the material and is useful in that two materials having the same MFI will, under the same conditions of processing, behave in the same way. The viscosity in poises can also be calculated from the experimental data by calculating the shear rate from the die dimensions and the rate of extrusion, and relating this to the pressure used.

(2) The plastic is extruded at a known rate (calculated from the rate of displacement of the plunger) and the pressure required for extrusion is measured. As before, the viscosity can be calculated and by repeating at different rates of extrusion the variation of viscosity with rate of shear can be plotted. By varying the temperature at which extrusion takes place the full information on the viscosity characteristics of the plastic melt can be obtained.

It will be appreciated that the melt flow index mentioned above does not tell

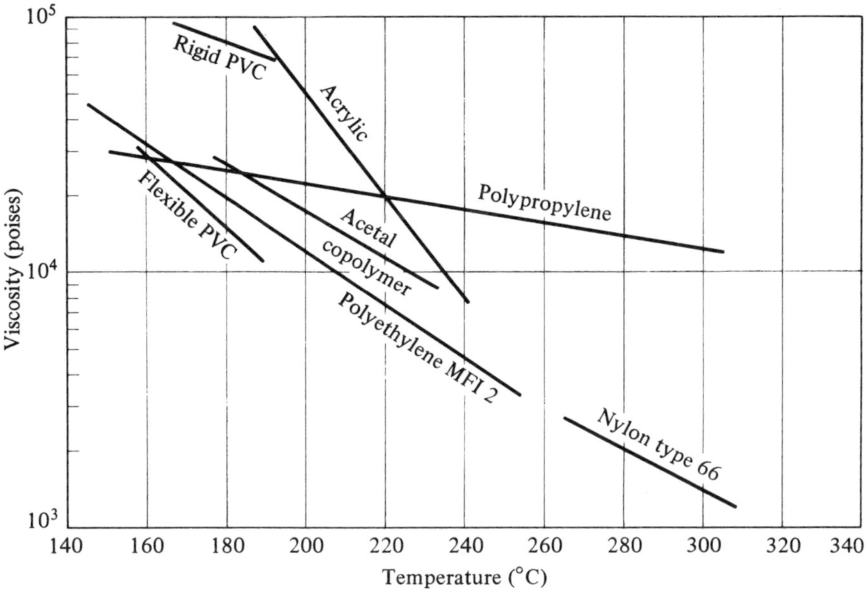

Fig. 4.9 Effect of temperature on viscosity of plastics (shear rate 10 s^{-1})

the whole story of the material's behaviour because it indicates the viscosity at only one rate of shear. However, the instrument used for the determination of melt flow index (sometimes known as a polythene grader or melt flow indexer because it was first used for characterizing polythene) can be used for determining another property of the plastic melt. The extrudate swells after leaving the die and by comparing the diameter of the extrudate with the diameter of the die the swelling ratio of the plastic (the after-shear swelling mentioned above) can be determined.

Figure 4.9 plots the effect of temperature on the viscosity of a number of plastics at constant shear rates. Figure 4.10 indicates the changes in viscosity that occur in a number of plastics at their normal moulding temperatures at various shear rates and Fig. 4.11 gives the swelling ratios that can be expected at different rates of shear.

In injection moulding, the plastic melt is being sheared whenever it is in motion. In the machine cylinder, shear rates of up to 1000 sec^{-1} may be experienced. Somewhat higher shear rates occur in the cylinder nozzle but generally the highest shear rate of all takes place in the gate into the mould. It follows, then, that in order to make best use of the pseudoplastic property of the plastic melt, the dimensions and geometry of the various parts of the feed channel to the mould need to be carefully calculated and the rate of injection accurately controlled.

Obviously, the lower the viscosity of the plastic melt, the less will be the pressure needed to convey the material from the cylinder to the mould and to fill the mould. Heating the material alone is not sufficient because, at low rates

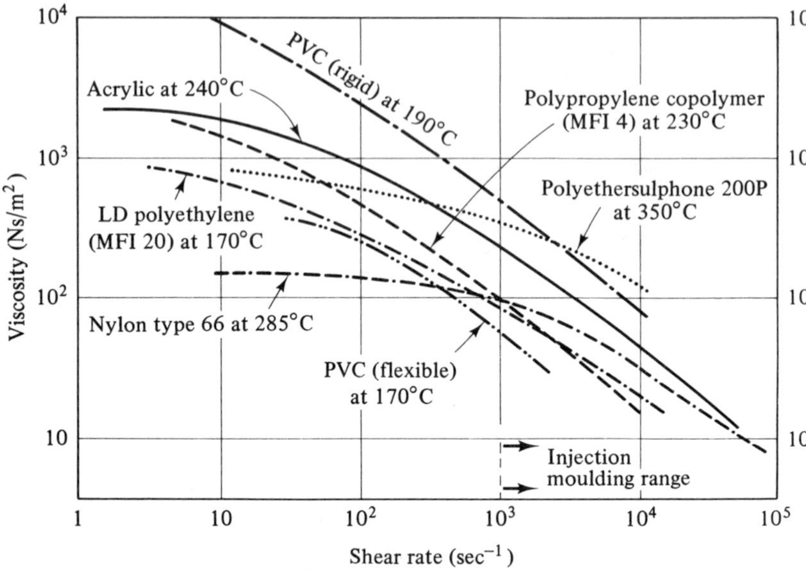

Fig. 4.10 Viscosity vs shear rate of plastics at their moulding temperature

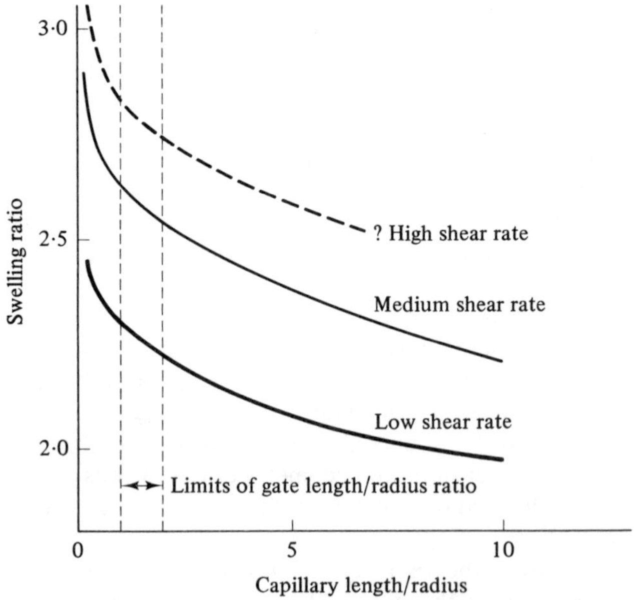

Fig. 4.11 Approximate after-shear swelling

of shear, even an easily moulded material, such as polythene, has a viscosity well above 5000 poise (the 'treacle' level) even at 200°C. To bring the viscosity down to between 5000 and 500 poise it is necessary to increase the shear rate with most materials to well above the 1000 s^{-1} level. To obtain the best results, particularly when making thin-walled mouldings, shear rates of 10^5 s^{-1} may be needed in the gate of the mould in order that the lowest possible material temperatures can be used.

What happens at shear rates greatly in excess of 10^5 s^{-1} is not fully known. The behaviour of plastic melts depends on two characteristics of the material which are in balance, namely the viscosity, which is non-recoverable deformation, and elasticity, which is recoverable deformation. One can, perhaps, envisage the molecules sliding against each other to give viscous flow reaching a limit when the molecules themselves are being extended. If this limit is reached at shear rates between 10^5 and 10^6 s^{-1}, which may indeed be the case, there would seem to be little point in further increasing injection rates in an attempt to shear the material at higher rates.

As the molecules slide against each other they tend to uncoil and to be drawn closer together. After the shearing has ended they coil up again and this effect—after-shear swelling—is greater if the period of shear is short, such as occurs in a short capillary. If sheared for a longer period it seems that the molecules not only uncoil but become disentangled and after-shear swelling becomes smaller. This observation would conform with the fact that as the disentanglement of the molecules becomes greater, the easier it is for the plastic to form into a crystal structure, which takes up less volume. It has also been observed that the time for complete relaxation of the sheared molecular structure may be quite long. Although the major proportion of after-shear swelling occurs in the first 0·2 s or less, if the material is prevented from 'freezing' it may take several minutes for the process to reach completion.

From the point of view of the practical moulder, the phenomenon of after-shear swelling offers a method whereby extra material can be injected into the mould to counteract the volume reduction that occurs as the moulding cools. If it is assumed that the sheared material will swell on relaxation by a factor of 1·5, and if the mould can be filled before the full amount of relaxation has occurred, up to 50 per cent of extra material can be injected into the mould. Since the total extra material required to compensate for the full reduction of volume on cooling is between 7 and 18 per cent, this rapid injection would seem to be the means of giving an adequate margin in the process. This, and other aspects of the use of rheological data in process control, and the diagnosis and cure of faults are discussed again in Chapter 6.

5

Moulds

Moulds for injection moulding are as varied in design, degree of complexity and size as are the components produced from them. In their simplest form, for use on the smallest machines, they may consist of two matching pieces of metal in which the impression of the article to be moulded is cut. A channel communicates with the nozzle of the plasticizing cylinder and the assembly is clamped in the correct position for moulding by means of a simple hand vice. In more complex forms, there will be means for ejecting the mouldings and cooling channels for controlling the temperature. There may be sliding side cores or forms, unscrewing devices for mouldings having a screw thread, and part of the mould may be kept hot to provide an extension to the nozzle so that sprue and runner systems remain fluid. Electrical heating may be installed and air is sometimes applied for ejecting the mouldings.

The traditional material for the construction of moulds is steel. Cavities and communicating channels are cut into blocks of good quality tool steel which can be hardened if necessary. Sprue bushes and guide pins and their bushes are made from very hard tool steel as are ejector pins and other ejection devices. The carrier, or bolster, of the mould is made of softer, mild steel. In fact, all the parts of the mould which do not have to withstand any frictional wear are best made from this grade of steel. Care must be taken in choosing the steel so that the grade matches the degree of finish required.

Because of the traditional use of steel, moulders have been slow to accept other materials, on the assumption that the abrasive action of the plastic passing into the mould would quickly wear away the surface polish. This is what happens with thermosetting materials and compression and transfer moulds are nearly always hardened and chromium plated. However, most thermoplastics have little abrasive action and for many components, even when runs of one million or more have been made, softer metals have been used. Among these are aluminium, brass, zinc, aluminium/zinc alloys (Kirksite) and low melting point alloys. For prototype work the last two of these materials are particularly useful because forms and cavities can easily be cast from them using plaster moulds. Also, for prototype work, metal-filled epoxide resins (for example, Devcon) have been successfully used and it has even been found possible to make one or two injection mouldings using a mould made of dental mechanics plaster.

When long runs are contemplated, the best possible mould should always be considered. Such moulds will be expensive but the cost per unit item when between ten and one hundred million items are to be made will be very small. A well-made mould will require little, if any, repairs over a long period but the cost will be high on account of the processes it has to undergo during manufacture. These are: annealing cycles to prevent distortion during use; hardening processes; polishing (up to 50 per cent of the total cost of a mould), and attention to fine detail, which are all expensive. Even before starting to make the mould a great deal of thought and the application of knowledge needs to be put into its design. It should always be remembered that is is easier and cheaper to alter a drawing than to alter a piece of machined steel. Second thoughts, so far as mould making is concerned, are not permissible when once the design has been approved and work started. Even if the design is changed, drawings should always be filed and catalogued so that they can be referred to at a later stage. This action may prove to be of value in the event of a dispute over a modification being followed by legal action. Verbal agreement on a matter so expensive as an injection mould could prove very costly if one party later disputed the agreement.

One of the banes of the custom moulder is the customer who asks for a prototype mould to be made to produce a small number of items for market evaluation and who comes back later with the request for many thousands of the mouldings made in the same mould. It must be understood that a prototype mould is exactly what its title implies. On some occasions a production run might be possible from such a mould but this should never be taken as a matter of course. Much can be learned from a prototype mould which probably makes the manufacture of a production mould easier and, therefore, cheaper, but this should be taken as the limit of its usefulness.

The first stage, then, in the production of an injection mould is the careful design of all the parts of the mould, consideration of whether component design has been optimized to suit injection moulding, and bringing the two concepts together in the making of the mould. This will then be proved and, if necessary, modified before handing over to the production department.

Mould design

Mould design is a very large subject which has been discussed in numerous books and publications. It is not intended in this book to give more than an outline with the object of making it more clear how problems and faults arise in injection moulding which can be related to the design and functioning of the mould. This design will depend on the article to be made and the machine on which the mould is to operate. The mould may be simple or complex, but there are some aspects of design that apply to all moulds, whatever their degree of

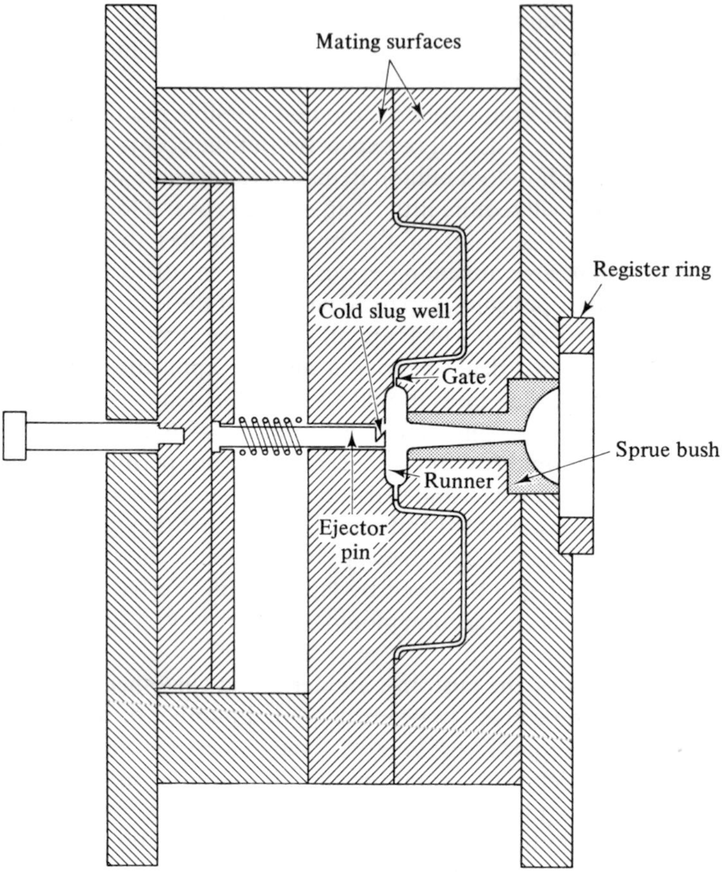

Fig. 5.1 A simple mould

complexity. These are: (1) Means of clamping or locking; (2) means of filling; (3) means of cooling; and (4) means for ejection of the moulding.

Before discussing these matters it is desirable to consider the basic parts of a simple mould (Fig. 5.1), so that the nomenclature of the various items is understood.

Material from the nozzle of the cylinder enters the sprue bush, which is a tapering hole cut in the part of the mould, which is fastened to the fixed platen of the machine. Often, at the end of the sprue bush, cut into the moving half of the mould is a small cavity in which an ejector pin slides. The first, slightly chilled, material passing through the sprue bush enters this cavity, which is the cold slug well, and solidifies. If the cavity has a slight undercut, or if the ejector pin has a Z-shaped undercut on its end, the cold slug acts as a sprue puller when the mould opens and pulls the solidified sprue out of the bush. When the mould is fully open the ejector pin moves forward allowing the cold slug and attached pieces of solid plastic to fall from the mould. Leading from the sprue bush, channels called 'runners' carry the material forward, through a restriction, known as the gate, into the mould cavity. Air trapped in the runners and cavity can escape at the parting line of the mould or mating surfaces, or along ejector pins. In fact, the air will escape from the mould through any gap which is not actually sealed off when the mould is closed. This general method of mould layout can be modified to suit other methods of clamping or any other matters which have to be taken into consideration.

Reverting now to the four basic aspects of design:

(1) *Means of clamping.* In the smallest machines, the mould may be completely demountable and when being filled is held in a simple vice. This can be vertically or horizontally acting to suit the cylinder, some of which are down-stroking and some horizontally acting. With a horizontal cylinder and a vertical clamp the runners and sprue bush are in the same plane and often, because the pressures involved are not very great, the hardened sprue bush is replaced by a simple runner cut into one or both halves of the mould.

In larger machines with horizontal clamping, thought should always be given as to whether a horizontal or a vertical flash line is either possible or desirable. In Fig. 5.2 a vertical flash line is shown, whereas Fig. 5.1 showed the more common horizontal flash line. With the vertical flash line, sometimes called a positive mould, it will be seen that material could not escape from the parting line of the mould until considerable opening movement had taken place. This means that if an over-full shot were to be made the mould would not 'flash', although the shot weight might be too great. It also has the

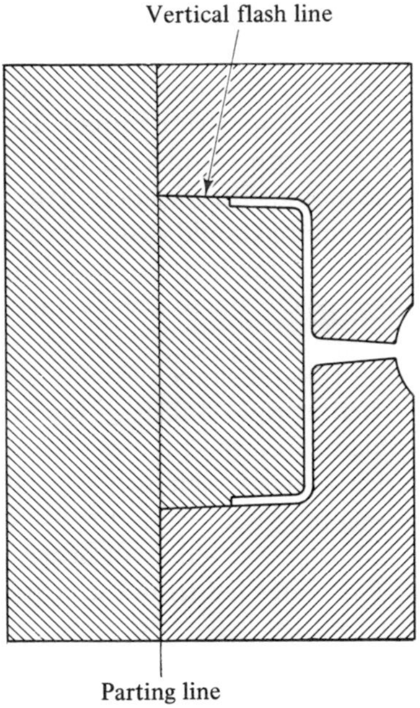

Fig. 5.2 Vertical flash (positive) mould

advantage that over-size mouldings can be made with a much smaller mould locking force. If the correct amount of material is injected into the mould it may open slightly (a few thousandths of an inch) but no material escapes and as the moulding cools the mould closes again thus 'compression moulding' the article.

(2) *Means of filling*. The way in which the mould is to be filled is determined by several factors, including the following: (a) The number of impressions (cavities); (b) the size and shape of the articles to be made; (c) whether the machine is to run automatically or not; (d) whether it is possible or desirable to mould without making waste material in the form of sprues and runners. Taking these in order: (a) the number of impressions to be used is determined by the output rate of components required, the size of machine available, and the degree of complexity of the mouldings.

With very complex mouldings made to a high degree of accuracy it is always wisest to consider using the smallest number of impressions to give the required output rate. This then becomes a matter of economics: will it be cheaper to use several small machines with, say, two or four impressions or one large machine with 24, 32 or even more impressions? One advantage of the former is obvious. Production does not completely stop if one four-impression mould out of eight moulds has to be repaired, whereas it does if the 32-impression mould breaks down. Unfortunately, not many custom-moulding shops are overburdened with numbers of small machines! It is also obvious that the more complicated the mould, the more elaborate will be the feed system connecting the various cavities. An increasingly elaborate mould makes cycling longer and sometimes the runners themselves determine how long the cycle needs to be.

An illustration of this concerns the moulding of automobile tyre valve caps, which were being made on a 40-impression mould. The cycle time was about 30–40 s because it took that length of time for the runners to become solid enough to remove from the mould. On considering the economics of the job it was found that the output required could easily be obtained on a single four-impression mould if the only cooling factor to be taken into account was that for the components themselves. Of course, the machine itself would need to be very fast cycling.

Turning now to (b): the size and shape of many articles, especially where core pins have to be inserted at right-angles to the opening direction of the mould, makes it impossible to feed the cavities except from a single, central runner. Simple, symmetrical articles such as bottle caps, are always best fed from a central gate and this may necessitate the use of a three-plate mould (q.v.). Circular articles fed from the side will nearly always show signs of non-circularity, whereas flat plates fed from the centre will often warp. It is seldom good practice to feed the cavity in more than one place unless a complete ring or diaphragm gate (q.v.) can be used. Large mouldings with integral holes are often best fed with a ring or diaphragm gate in the largest hole or in the hole nearest to the centre of symmetry.

With regard to (c), if a machine is to run automatically it is desirable that the mouldings are automatically stripped from the runners so that mouldings and waste can be separately collected. This may require 'tunnel' or 'submarine' gating (q.v.) or pin point gating in a three-plate mould. In any case, the gate must be small enough to break easily to allow ejection of mouldings and waste. Of course, larger mouldings or even sprays of mouldings can be produced automatically without automatic de-gating, in which case other gating systems can be used.

The fourth criterion for filling, (d) raises the question of so-called sprueless moulding. There are several types which necessitate some description in order to identify the most suitable method and this will be given later in the chapter.

(3) *Means of cooling*. One of the most important aspects of mould design is the provision of suitable and adequate cooling arrangements. In all injection moulding, even though it may involve having a heated mould, the net purpose of the mould is to cool the molten plastic. If a mould had no means of cooling and was insulated to prevent any escape of heat by conduction, convection, or radiation, it would quickly reach the temperature of the material being moulded and would no longer fulfil its function. The amount of heat which has to be removed can be calculated from the weight of the article and its enthalpy at moulding temperature, allowing for the fact that the temperature of the moulding when ejected will generally be well above ambient. This means that it is possible to estimate the amount of cooling required, the dimensions of the cooling channels, and the flow of coolant.

To take, as an example, the moulding of an article weighing 100 g in polypropylene: it is found in practice that about one-half of the total heat in the moulding is taken away in the mould, so with a figure of 130 cal/g for the moulding enthalpy, each moulding will lose 6500 cal to the mould. At an output rate of three mouldings per minute the total heat to be removed in 1 h is 1 170 000 cal. If the cooling water rises in temperature by 5°C, a total flow of 234 l/h or about 0·86 gal/min is required.

Assuming that the moulding is in the form of a shallow box with a hinged lid, some 25 cm in length and of total width (box and lid) 15 cm, the layout could be as indicated in section in Fig. 5.3. The surface of the mould should never exceed 140°C (HDT of polypropylene) and will probably average

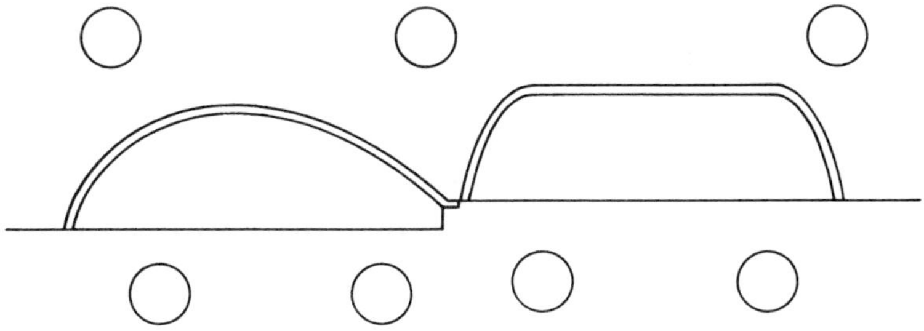

Fig. 5.3 Possible arrangement of cooling channels for box with hinged lid

100°C. The coolant in the mould should be kept at 60°C and 325 cal/s of heat must be removed. If seven cooling channels each of 1 cm diameter are bored through the length of the mould in the approximate position shown, the average thickness of steel through which heat would be conducted would be about 4 cm (thermal conductivity of steel 0·115 cal/cm s °C). It can also be assumed that only half of the area of the cooling channels would effectively be conducting the heat and that the coolant would be removing 100 per cent of the heat conducted through the steel. The total length of cooling channel needed is given by:

$$\text{length of channel (cm)} = \frac{\text{total heat to be removed} \times \text{thickness of steel}}{\text{thermal cond. of steel} \times \text{temp. diff.} \times \text{area}}$$

$$= \frac{325 \times 4{\cdot}0}{0{\cdot}115 \times 40 \times 1{\cdot}57} = 180 \text{ cm.}$$

that is, the effective cooling length of each of the seven channels is about 26 cm.

The flow of coolant should always be turbulent through the cooling channels in order to obtain the best heat transfer. This is checked by calculating the Reynolds number, a dimensionless figure which is used in flow calculations and which, if above 3000, shows that the flow is turbulent. It is given by the equation:

$$\text{Reynolds number} = \frac{98{\cdot}285 \times Q}{u \times d}$$

where Q is the flow rate in cm^3/s
u is the viscosity of the liquid in centistokes (water = 1)
d is the diameter of the pipe in centimetres

In the example above, $Q = 85$ cm^3/s, and $d = 1{\cdot}0$ cm

$$\text{Reynolds number} = \frac{98{\cdot}285 \times 85}{1 \times 1} = 8345$$

and the flow is therefore turbulent.

This is, of course, only a very rough approximation but it serves to show that it is possible to obtain some indication at the outset of the design work of the dimensions of the cooling channels. Their exact location will depend on other factors such as the position of ejector mechanisms, runners, etc., but it is best to decide on cooling requirements at an early stage rather than to add the cooling channels when the mould is already made and, possibly, hardened. Further discussion on the positions of cooling channels will appear later in the chapter when considering the effect of cooling arrangements on warping. There is, however, one more important reminder at this stage: the sprue and runner system may contribute quite a large proportion of the total heat to the mould and their effect cannot be ignored.

(4) *Means of ejecting the mouldings.* Some mouldings require little in the

way of ejectors: others need very complicated systems, but in all cases ejection should be firm, yet gentle. Ejector pins which puncture the surface of the moulding so that it remains impaled upon them instead of falling from the mould are worse than useless. The first requirement, then, is that ejectors should act on parts of the moulding which become solid quite quickly and this means that cooling channels should be located fairly close by. Thin sections require more careful treatment. They often remain on the male portion of the mould as it opens and a sleeve stripper plate can be used instead of pins. Another useful type of ejector is the valve ejector, which allows air to enter a hollow moulding such as a bowl or beaker and causes it to be ejected. The opening of the valve as the mould opens can be made to coincide with a blast of air which will blow the moulding off the form. Care is necessary or the moulding is blown straight back into the cavity!

Ejectors should be designed to do their work at a time when the mould is moving slowly and, for this reason, hydraulic ejector systems programmed to operate at the correct moment are preferred. They can be made to operate gently even though the mould platen is moving rapidly.

Moulds are normally vented at mating surfaces and at ejector pins; in fact wherever there is a part of the mould sliding or moving against another part there is likely to be venting. With rapid mould filling, air venting may cause a considerable amount of heat to be developed at small gaps and care must be taken to make sure that hardened surfaces are not damaged by oxidation. For this reason, the provision of proper venting arrangements should be an integral part of mould design and air escape should not be left to chance.

Mould-making methods

A complete account of the very many methods of mould making would be out of place in a book devoted to the causes and cure of faults in injection moulding. There are several excellent books on the subject to which reference can be made. However, to assist in the understanding of how and why problems occur in connection with the mould, a brief sketch of the main mould-making methods involved will be included.

The chief process used is machining, in which the shape to be generated is cut on a lathe or milling machine to the required dimensions. Cutting of very hard steels is difficult so softer steels are used and subsequently hardened or nitrided. During hardening, the heat treatment which the metal receives may cause stresses which were present in the metal block to be relieved and slight distortion may result. The use of normalized steels help to reduce this, but a number of annealing treatments as the machining proceeds may be necessary with very accurate work. Most metals can be machined and for simple circular shapes the lathe is the most useful tool.

A modification of machining is die-sinking, in which a negative pattern is used which a pantograph tool will follow to generate the appropriate shape. Great accuracy is possible as the pantograph can be arranged to give a reduction of dimension of as much as 10:1 compared with the pattern. Engraving also uses a pantograph and is chiefly used for lettering.

Where repetitive work is done, numerically controlled (N-C) machinery is used. It is expensive and the work load needs to be considerable with a large saving of labour before its use can be justified.

Very hard steels can be worked by either electrochemical machining (ECM) or electric-discharge machining (EDM), often referred to as spark erosion. ECM is very rapid and 1 in^3/min of metal can be removed each minute for each 10 000 A of current, which is obviously at very low voltage. Surface finishes are good but electrodes, which are of suitable shape to generate the form required, are expensive. EDM is generally used for smaller jobs and is a substitute for die sinking. Blind holes are quickly generated and sharp angles are possible, dimensions better than 0·001 in being easily obtained. Newest machines employing transistorized pulse generators which produce very rapid sparking give little electrode wear and a good surface finish. Electrodes are reasonably easy and inexpensive to make, graphite and copper being preferred materials.

These methods of mould making involve cutting metal away from a form. Two other methods in use involve the reverse, namely, building up metal to give the desired shape. Electroforming is a process in which metal is electrically deposited on a reverse model of the shape to be formed and the shell of metal thus produced is backed with a fusible casting alloy such as Hoyt metal or Kirksite. Cooling coils of soft copper pipe can be incorporated into the alloy backing. In metal spraying a low melting point alloy is flame sprayed on to a positive model and the shell so formed is backed with a similar alloy. Because the alloys used in both these processes are softer than desired the forms are mounted in a steel bolster and the locking force is taken up on hardened steel pads incorporated in the bolster.

Cast metal moulds of aluminium, zinc, Kirksite and beryllium copper are also used satisfactorily in injection moulding. In common with all metals, their thermal conductivity is relatively good (better than 0·1 cal/cm s °C, Be/Cu being about 0·5 cal/cm s°C) and even with those alloys of very low melting point (c. 160°C) there is no danger of the hot plastic melting the mould. However, care must be taken to minimize flashing of the mould because this is always accompanied by excessive air-venting with its consequent production of heat.

Gates for injection moulds

The gate, that is, the constriction between the feed system and the mould cavity, serves several purposes. It freezes rapidly and prevents material either entering or leaving the mould; it provides an easy means of separating mouldings from the feed system and it provides a small portion of the feed system where very rapid shearing of the material occurs. There is no doubt that incorrect gating is one of the most frequent causes of faults in injection mouldings.

(a) *Side gate.* Where a moulding is fed on the side from a runner system a restricted gate may be used. It is important with this type of gate to ensure that the jet of material entering the cavity is broken up within a short distance (say, $\frac{1}{4}$ in or 6 mm) of the entry point. If this does not occur, the jet travels across the free space, becomes cooled, and fails to fuse properly together to form a homogeneous mass. This fault, known as jetting, is avoided by causing the jet to impinge on a pin or mould surface within 6 mm of the entry point. If there is no suitable pin or other surface, the gate should be angled near a corner of the cavity so that it impinges on one of the walls. If possible, a side gate should be round, or nearly so. If the runner is fully round, the gate can be cut in each half of the mould but if the runner is trapezoidal the gate should be of similar configuration but, of course, much smaller. A tunnel gate (or submarine gate) is a special form of side gate in which the entry to the cavity is through the side wall but below the level of the parting line. When the mould opens, the sprue and runner is sheared from the moulding as this is ejected (self de-gating).

(b) *Fan gate.* This is really a restricted gate which is feeding through a slit rather than through a hole. The runner should be fanned out to the full width of the slit at the same general dimension as the remainder of the runner and the land length of the slit should be as short as possible, commensurate with giving enough strength of metal between runner and cavity.

(c) *Flash gate.* This gate feeds through a slit which extends virtually along one side of the moulding. The runner should extend beyond the slit at each end by a few millimetres so that the material will then completely fill the runner and tend to break through the slit in one wave. With a full round runner, the gate can be made to feed at the parting line of the cavity, supposing that this is cut in both sides of the mould, and this avoids giving an undue shearing force on one side of the web of material entering the mould. If the cavity is cut in only one half of the mould and fed from a trapezoidal runner, the same effect can be avoided by having an insert or raised portion of the opposite half of the mould to direct the flow of material into the centre of the moulding. However, this means having a rather thin and therefore easily damaged portion of metal on the fixed half of the mould.

(d) *Diaphragm gate.* When making cylindrical articles or feeding into a hole in a moulding, the runner system is extended at its general dimension to almost fill the hole, leaving only a narrow section through which a slit allows material to enter the mould. With regard to strength of the metal, the conditions as applied to fan gates should be observed.

(e) *Ring gate.* External, internal, or vertical ring gates are used for feeding cylindrical mouldings and as the name implies, the runner circles the cavity at its normal dimensions. Used instead of a diaphragm gate, there is a saving of material in the runner.

(f) *Pin-point or pin gate.* Here the gate is cut through a plate of the mould to provide the point of entry for the material. It can be fed from a runner system cut into the fixed part of the mould and restrained by a floating plate (*see* Three-plate mould) which is in the centre of the mould and carries part of the cavity, or from an extended nozzle or hot runner system.

(g) *Sprue gate.* A sprue gate is often used for making articles such as buckets and bowls. Sprue gates are nearly always too large and the incorrect amount of shear is developed in the material as its enters the mould, allowing excessive mould packing and the consequent failure in service of the moulding.

Many of the points mentioned above are illustrated in Fig. 5.4.

Calculation of gate sizes

Gates should always be made small at the start; they can easily be made bigger but cannot so easily be reduced in size. The question is: how small? Gate dimensions are important. Because the pressure drop in a system is proportional to the length of the channel, the land length of the gate should be as short as possible but, as already mentioned, the strength of the metal may be a limiting factor. On average, 0·040–0·060 in (1·0–1·5 mm) is a suitable length.

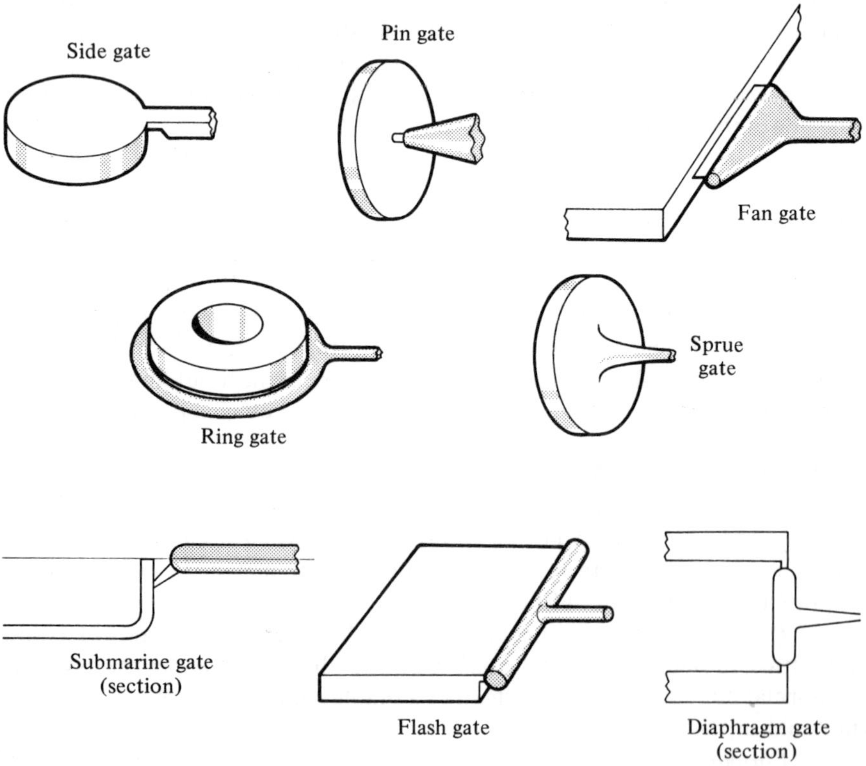

Fig. 5.4 Gates

It will have been seen from the discussion on gates that they are of two main forms—holes and slits (a ring gate or a diaphragm gate can be thought of as a slit of length πd). The hole is preferably round. For reasons already discussed it is desirable to think in terms of shearing the material entering the mould at a maximum shear rate of $10^5\ s^{-1}$ and the dimensions of the gate are calculated to give this shear rate.

The Newtonian shear rate for a circular die is given by the formula:

$$\dot{\gamma} = \frac{4Q}{\pi r^3}$$

where $\dot{\gamma}$ is the shear rate
Q is the volumetric flow rate in $cm^3\ s^{-1}$
r is the radius of the die in centimetres.

For a slit:

$$\dot{\gamma} = \frac{6Q}{lt^2}$$

where l is the length of the slit in centimetres
t is the width (or thickness) of the slit in centimetres.

Equating these formulae to $\dot{\gamma} = 100\,000\ s^{-1}$ to give r and t respectively:
For the round hole

$$r = \sqrt[3]{\frac{4Q}{314\,000}}$$

and for the slit

$$t = \sqrt[2]{\frac{6Q}{100\,000l}}$$

Thus, for a volumetric flow rate of $100\ cm^3\ s^{-1}$ the gate sizes will be:

For the round hole

$$r = \sqrt[3]{\frac{400}{314\,000}} = 0{\cdot}108\ \text{cm}\ (0{\cdot}040\ \text{in})$$

and for the slit of length 10 cm (4 in)

$$t = \sqrt[2]{\frac{600}{1\,000\,000}} = 0{\cdot}02\ \text{cm}\ (0{\cdot}01\ \text{in})$$

Figures 5.5 and 5.6 indicate gate dimensions for various filling and shear rates for round holes and slits.

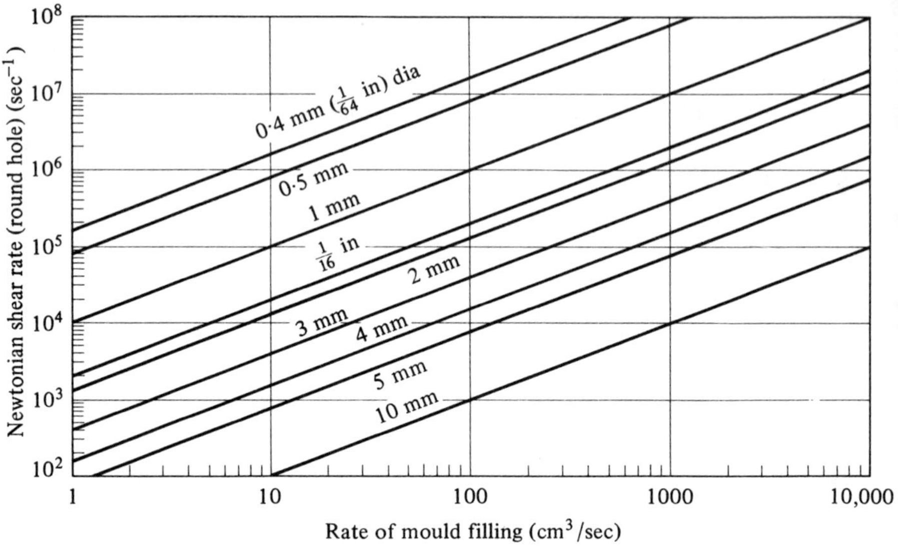

Fig. 5.5 Gate sizes for various filling and shear rates (circular gates)

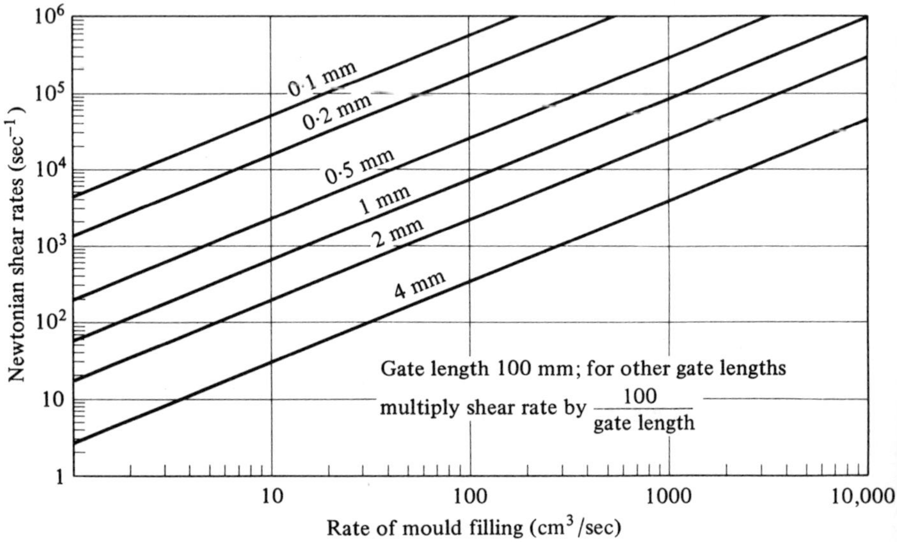

Fig. 5.6 Gate sizes for various filling and shear rates (slit gates)

Runner systems for injection moulding

The standard mould as illustrated in Fig. 5.1 has full round runners, and from the point of view of ease of mould filling these are the best. They give the lowest pressure drop per unit length and should be used if possible. They also give the best configuration for gate entry. However, they are more expensive to make because they have to be cut into each half of the mould and a compromise shape is used. This is a trapezoid which can be cut into only one side of the mould and again, to achieve the lowest pressure drop characteristics, a very generous radius should be given to the corners in the base of the cut. In fact, if the trapezoid can be cut so that it would exactly accommodate a fully round runner of the desired diameter with the sides of the runner tapered at 5° from vertical above the half way line, this is almost as good as the round runner. This is the configuration used in calculating the weights of runner systems given at the end of this section (Fig. 5.7).

Where several impressions are to be served from a runner system, it is always best to arrange that the distance from sprue to gate is the same for all impressions. This is fairly easy to achieve with 2, 4, 8, 16 etc. impressions but needs careful planning for 3, 6, 12, and 24 impressions. If the flow path distance varies, there is the possibility of some cavities filling preferentially or of some of the mouldings having slightly different dimensions. Two opposing factors are present: the material reaching the farthest distant cavity will be slightly cooler and will therefore shrink a little less than the hotter material in the nearer cavity. However, the pressure drop in the runner to the more distant cavity will be greater than that to the nearer cavity and will slightly compensate for the lower shrinkage, but if the mould is filled very rapidly this will have only a marginal effect. In practice, the farthest cavity, if it fills at all, will give the largest moulding.

Failure of some cavities to fill indicates that the runner system is unbalanced and, assuming that all the runner-lengths are the same, it means that the gates of those cavities are giving too great a pressure drop. This can be reduced either by opening the gate or by shortening the land length. The former alters both shear rate and pressure drop; the latter only reduces pressure drop and has the virtue of being more accurate. For example, a gate may only require a

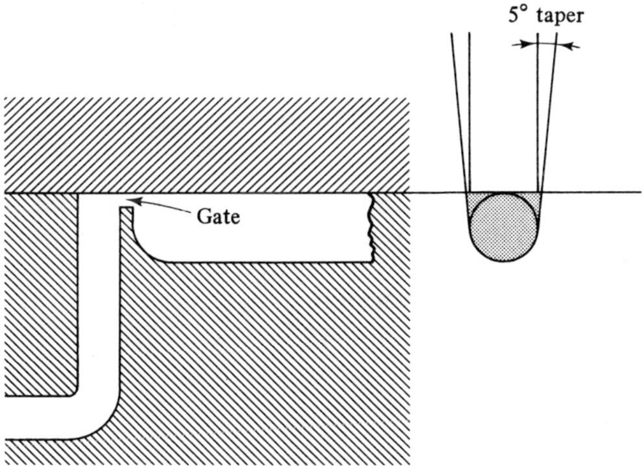

Fig. 5.7 Rounded trapezoidal runner

very slight adjustment to enable the cavity to fill and on a land length of 0·060 in (1·5 mm) removal of 0·002 in (0·05 mm) of metal will give a reduction in pressure drop of about 3 per cent. Removal of the same amount of metal on the diameter of a round gate will alter the pressure drop by 12 or 13 per cent.

'Family' moulds

It is not good practice (although it is very often done) to have greatly differing sized cavities in a single mould. Many so-called family moulds require this and almost without exception cause difficulties. It may be that some of the mouldings warp, others show excessive frozen-in strain and any attempt at balancing the cavities leads to long and tedious trials. The practice is very common in the United States of America and is often advocated in the United Kingdom, particularly in U.S.-owned companies. One instance was observed of a family mould for making a child's toy consisting of a set of eight beakers, each fitting inside the other. Only occasionally was it possible to produce a complete, fault-free set of mouldings. Unfortunately, the loss of one component really meant scrapping the whole spray. Perhaps this is not so important where very long runs are common, but getting this mould right was very difficult. Producing scrap mouldings is very expensive and, although the material is all recovered and re-processed, the cost of doing so is not insignificant and it makes the finished article much dearer to produce.

Another family mould for a model truck gave similar problems. If one cavity filled, another flashed and in a set of four rods which needed to be straight, some came out warped. In this mould, most of the cavities had similar shot weights but they had very different shapes and dimensions. Balancing the gates and runners was an almost impossible task.

In all such projects, this problem should be considered at the outset because it may be found more expedient to break the job down into smaller units and to use smaller machines and faster cycling, with the bonus of being able to run one or more of the moulds for a longer time to make up deficiencies without any waste.

Runner dimensions

Runner dimensions have to be decided by balancing two factors: (1) the cost, in material and cooling-time, of large runners and (2) the need to minimize pressure losses. Table 5.1 gives the weights per inch and per centimetre in grammes for round and trapezoidal runners of various sizes, allowing a 5° taper per side on the trapezoidal shape (*see* Fig. 5.7).

Three-plate and hot runner moulds

As implied by the name a three-plate mould has an additional plate which floats between the fixed and moving plates. The sprue and runner system is

Table 5.1 Weights of runner systems in grammes for material density 1 g/cm³

Diameter (in)	Weight per inch		Diameter (mm)	Weight per centimetre	
	Round	Trapezoidal		Round	Trapezoidal
$\frac{1}{8}$	0·201	0·234	2	0·0314	0·0366
$\frac{3}{16}$	0·452	0·527	3	0·707	0·0823
$\frac{1}{4}$	0·804	0·937	4	0·1256	0·1463
$\frac{5}{16}$	1·256	1·464	5	0·1963	0·2286
$\frac{3}{8}$	1·809	2·107	6	0·2826	0·3292
$\frac{7}{16}$	2·463	2·868	9	0·6359	0·7407
$\frac{1}{2}$	3·217	3·750	12	1·1304	1·3167
$\frac{5}{8}$	5·026	5·854	15	1·7663	2·0574
$\frac{3}{4}$	7·237	8·430	18	2·5434	2·9626

For other materials, multiply the above figures by the relative density.

contained between the fixed plate and the floating plate and the runners communicate with the cavities of the mould through pin-point gates carried on the ends of secondary sprues (sometimes known as carrots), cut into the floating plate. The cores are usually on the moving plate and the cavities in the floating plate. Ejection of the components from the cores is by the usual means but special arrangements have to be made for ejecting the sprues and runners. A cold slug with a slight undercut in the floating plate provides sprue pulling for the main sprue and small undercut pins held by springs in the fixed plate withdraw the 'carrots' from their bushes and break off the components at the gate. When the mould is fully open these pins are brought forward by an ejector bar attached to the floating plate and this ejects the runner system from the mould. A typical arrangement is shown in Fig. 5.8.

Any operation in which large sprues and runners have to be used is not so economical as it could be and, whenever possible, sprueless moulding should be carried out. Two main methods are used, each involving the use of a modified three-plate mould. These are (a) hot runner moulding and (b) insulated runner moulding.

(a) *Hot runner moulding.* The mould used is a modified three-plate mould in which all the channels—main sprue, runners and carrots—are kept molten. The main sprue may be replaced by an extension on the nozzle and the runner system is somewhat larger than usual and is heated with rod heaters. Alternatively, it may take the form of a heated manifold. It is important to insulate the heated part of the mould from the part carrying the cavities, either by means of air gaps or by Sindanyo pads, and to allow for differential expansion in the mould parts, which could be as much as 0·002 in/in (0·002 cm/cm).

(b) *Insulated runner moulding.* This system relies on the very low thermal conductivity of the plastic allowing a molten core to remain in the centre of an oversize runner during the time which it takes to open the mould, extract the moulding, and close it again. This runner is generally fully round and cut into

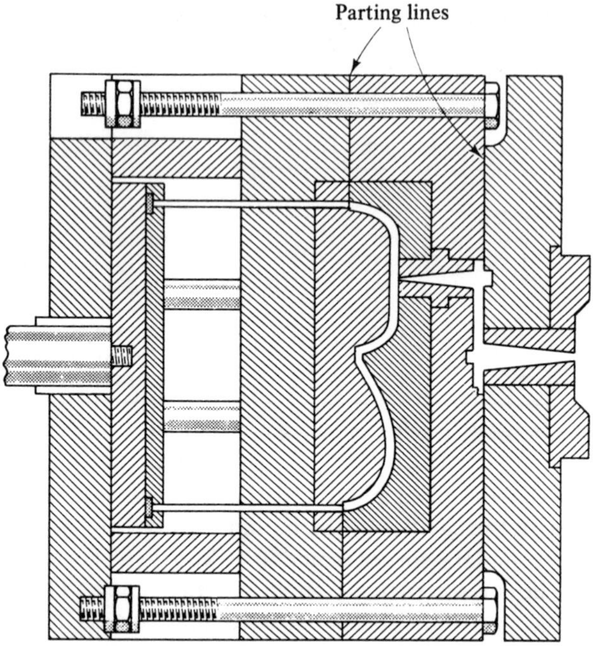

Fig. 5.8 3-plate mould

both fixed and floating plates of the mould. An extended nozzle, instead of a normal sprue, feeds the runner and the carrots are kept molten by means of cartridge heaters. If the runner sets up and does not allow material to pass along it, the fixed and floating plates are separated and the runner is extracted. Normally, the fixed and floating plates are held together by means of a latch. Insulated runner moulds will operate fully automatically for very long periods and they have the advantage that there does not need to be such a large temperature differential between the two portions of the mould as is necessary in the heated runner system. In starting up, the feed has to be adjusted so that the runner will completely fill and this may present difficulties if the shot weight without the runner is close to the maximum for the machine.

Extended nozzles

To dispense with the sprue when using hot or insulated runner moulds or to feed direct into the mould cavity, extended nozzles are used. They are suitable for single-impression work or, in the form of a manifold nozzle, for multi-impression moulding. Not more than four such nozzles on a manifold would normally be recommended. Two basic types known simply as 'the extended nozzle' and 'the Italian sprue' are used. In the former, the nozzle extension has a small flat on the end, which forms part of the base of the moulding, and to prevent it from freezing the nozzle is heated with a small cartridge heater.

Alternatively, the extension may be made of beryllium–copper which has about five times the thermal conductivity of steel. An enlarged water-cooled sprue bush takes up the pressure of the nozzle and the contact area is kept as small as possible. An air gap between the nozzle and the sprue bush serves to insulate it from the cooler mould. The Italian sprue also engages with a cooled sprue bush, but a plug of solid plastic which forms at the end of the nozzle constitutes about one third of the sprue. The core of this remains fluid and under the high injection pressure the stream of hot material can break through and enter the mould. An under-cut on the end of the nozzle allows withdrawal of the solid plug should the core freeze.

Mould cooling systems

The most effective cooling method for any mould would be to pump the coolant through the cooling channels counter-current to the flow of molten plastic into the mould. This method is adopted for pressure die casting and by using it the temperature difference between coolant and mould is kept as great as possible, on the assumption that the coolant increases in temperature as it progresses through the cooling channels and the material decreases in temperature as it penetrates into the mould. In the moulding of plastics, this method is suitable only for the moulding of thin-walled containers where quantity is of more significance than quality.

Where there is the possibility of differential shrinkage and the consequent warping of mouldings, attention to the direction in which the coolant flows through the cooling channels can help to alleviate, if not altogether prevent, warpage. The general principle is to make the coolant flow through its channels *in the same direction* as the material flows into the mould. This is less efficient in cooling effect but gives better quality mouldings, although at a lower output rate.

Differential shrinkage and cooling effects

In the moulding of a bowl or bucket, or even of a centre-fed flat plate, it is often found that radial shrinkage will be double that in the circumferential direction. Although, as indicated earlier, this may be caused or made worse by indifferent moulding techniques or by machine faults, it is not always possible to prevent it from occurring, and it is then that differential cooling is useful. The first material to enter the mould is under very little pressure as it flows into the mould channels but, as it proceeds towards the extremities of the mould, it is progressively cooled. The following material is hotter and it begins to pressurize that already there; but as the mould has filled, a skin of solid plastic has formed on each wall (*see* Chapter 6) which has reduced the gap through which the material passes and so increased the shear rate. The extra shear stress causes orientation of the molecular structure and the melt is generally too viscous to allow for relaxation of all the stresses which are developed. The result is that the shrinkage of the material becomes much greater in the direction of flow and orientation; that is, in the radial direction, compared with that at right-angles to it, the circumferential direction.

Controlling the temperature of the mould will not completely prevent the above series of events from happening, but it can help to even out the effects. Cooling to the maximum at the point of entry of material into the mould reduces the time needed for cooling in this area which, in turn, reduces the amount of crystallinity in the solid material and decreases the size of the crystals. This reduces the amount of after-moulding shrinkage. The perimeter of the mould, cooled less quickly because the coolant is at a slightly higher temperature, has the opposite effect. There is also a slight annealing effect which allows a little molecular relaxation and so relieves some of the effects of orientation. To achieve the best results, therefore, all cooling arrangements should ensure that the maximum cooling effect is nearest to the point of entry into the mould. Thereafter, the cooling effect should be as even as possible across the mould, although with large mouldings, it may be necessary to have two, or even three or more cooling systems, each run at a different temperature.

It is sometimes found that the walls of a box-shaped moulding will tend to bow inwards and that this effect can be alleviated by keeping the core cooler than the cavity. This would appear to contradict what has already been said, because bowing always takes place where there is thicker or hotter material. If one considers how a box mould fills from a centre gate in the base, it will be realized that the flow of material is radial in the base of the box and in virtually straight lines up the walls, giving a great deal of shear in that direction. The major shrinkage will therefore be in that direction and the perimeters of the walls end up bigger than they should be for the size of the base. The restraining influence of the base causes them to bow inwards. If the core is then cooled more than the cavity, the material on the inside of the walls will shrink less than that on the outside and the net result will be that the walls become straight. If the core is cooled too much there will be a tendency for the walls to bow outwards.

Component design

There are several basic concepts in component design which, if followed, will help to relieve many of the problems that beset the injection moulder. Among these are:

(a) Maintain the wall thickness as uniform as possible. Differences in thickness of only 15–25 per cent will often lead to sink marks forming in the thicker parts. At worst, voids could be produced and these could affect the performance. Rigidity can often be imparted by introducing a stepped section (*cf.* corrugated roofing sheets) and even box sections are made more rigid while keeping their thickness even, by slightly rounding the corners. If ribs and bosses are to be introduced, sink marks may occur and may be difficult to avoid. A useful rule is to make ribs only two-thirds of the general wall thickness and to design bosses similarly, making holes to penetrate one-third of the depth into the wall.

(b) The wall thickness should be no greater than is absolutely necessary to give the strength and rigidity needed. This keeps the use of material to the minimum and gives the shortest possible time cycle. Problems of mould filling can probably be solved by careful use of multiple gating.

(c) Undercuts and re-entrant shapes cause extra costs both in mould production and in moulding time (side core operation very often increases the time cycle). They can often be avoided by the use of simple assembly methods applied to two or more components. One of the most favoured of these is the interference fit in which one component has a clip or other device which engages strongly with the other component and holds the two together (*cf.* the poppet bead). Welding by friction, hot plate, or ultrasonic means can also be used, but in general, cementing with solvent or solvent-based adhesives is not recommended for the custom moulder because of fire and toxicity hazards and problems of solvent recovery.

(d) Avoid the use of moulded-in inserts if possible. Use instead staked or spring-clip inserts into blind holes which are easier to make by injection moulding. Insert moulding is often very wasteful of time, of both operator and machine, and this is generally the most expensive part of the process.

(e) Give the most generous radii possible. Sharp corners, both external and internal, are natural stress-raisers but a radius of 0·060 in (1·5 mm) will greatly relieve the situation and may increase impact strength by as much as five or ten times.

Component designers may call for something that is difficult to mould. As the producer of a well-known brand of hand-tool manufacturers put it in an advertisement: 'Occasionally, so rarely, in fact, that one feels rather guilty about mentioning it, designers have been known to create things which are just a teeny-weeny bit difficult to make. Production men rejoice in challenges of this kind, of course, and the English language would be poorer without them. . .' The same could, perhaps, be said of the injection moulder.

Designers may be working in a state of ignorance—ignorance of the service conditions of a component, ignorance of the service performance of a material, and ignorance of the many factors involved in proceeding from component design to complete plastic moulding. There is a need for education but, even more, there is a need for consultation and cooperation. Regular liaison between end user, component designer, mould designer, and moulding technologist will prevent many of these problems from arising.

6

The process and its control

Successful injection moulding depends on the ability of the machine to operate under the correct conditions of pressure, temperature, and speed of operation for the component being moulded and the material being used. It also depends on the ability to control the conditions of operation within close limits for long periods and on the exact repetition of the cycle of operations many thousands, perhaps millions, of times. These requirements can be considered first in terms of whether the machine is basically suited to the job, and then as to whether it can be controlled to the limits required.

Determination of whether the machine is capable of meeting the demands to be made upon it can be a matter of trial and error, but if some fairly straightforward calculations are made it is often possible to decide on the feasibility of a job without going any further. Some assumptions need to be made and accepted because a full mathematical analysis of the injection moulding process is extremely complex and is not the kind of exercise which the average practical moulder would wish to undertake. If the assumptions are accepted, it becomes possible to arrive at simple mathematical approximations which will give results close enough for the purpose required. The type of information needed is:

(a) What injection rate is required for satisfactory mould filling? In other words, how long is the injection time to be?
(b) Assuming the injection rate in (a), what will be the pressure requirements?
(c) With the pressure requirements of (b) and the injection rate of (a), what amount of hydraulic fluid will be required for actuating the machine, and how big a pump motor will be required to give the rate of flow at the pressure needed?

Having decided on these matters, the question as to whether the machine can be controlled accurately over the required number of cycles and at the desired rate becomes a matter of machine mechanics.

Determination of injection rate

The experienced moulder, looking at a moulding, would be able to make a reasonable estimate of the time cycle for moulding, including the time required for injection. He would assume, in general, that the most rapid injection rate possible would be beneficial and he would equate this to his knowledge of the capabilities of the machines in his moulding shop. Some estimators work on this principle, too, and this may account for some of the large discrepancies in cost estimates between one moulding organization and another. This approach can, at best, be inspired guess-work and a more rational approach is needed. It is possible to calculate, with a fair degree of accuracy, the various portions of the cycle and one of these is the injection rate.

In making the calculation, it is assumed that, as the material enters the mould cavity it begins to solidify on the walls of the mould, reducing the gap through which the following material flows. It has been observed by I. T. Barrie ('Understanding How an Injection Mould Fills', *Plastics Engineering*, Society of Plastics Engineers Inc., Aug. 1971) that the layer of solidified material is uniform at all points at the instant the melt front reaches the cavity extremities and that the thickness of the solidified material is proportional to the filling time of the mould. This latter observation can be expressed mathematically:

$$d = Ct^{1/3}$$

where d is the thickness of the layer in millimetres
t is the time in seconds
C is a constant, known as the freeze-off constant

It appears that C is related to the dimensionless factor

$$\frac{T_x - T_m}{T_c - T_m}$$

which was discussed in Chapter 4. For a material such as polypropylene the constant C is about 0·3.

The formula gives the thickness of the layer on only one wall of the mould. Since both walls are involved in closing the gap between them, to give the time for mould filling the formula is modified and transposed to become:

$$t = \frac{d_w^3}{0{\cdot}216}$$

where t is the time of filling in seconds and d_w is the minimum wall thickness of the moulding in millimetres.

Obviously, this formula will only give a *maximum* time for mould filling, because it is really giving a means of calculating the time when the gap between the walls of the mould reaches zero by solidification of the plastic. At this stage no more material will pass further into the mould except under extremely high pressure, with its attendant development of high shear stresses. A further assumption, therefore, needs to be made and this depends to some extent on observation and practical common sense. The question needs to be answered, 'at what stage in the solidification process does the mould become virtually un-fillable?'

A very rough approximation to the answer is to assume that the mould should be filled before half of the available gap has been taken up with solidified material, but perhaps a more accurate result can be obtained from published information on the material being used. Most manufacturers give a maximum flow-path ratio for their materials, for example, 200:1. This means that with a wall thickness of 1 mm, the molten plastic would flow to a distance of 200 mm.

Suppose that in a given moulding the material needs to flow for 100 mm, the minimum allowable gap will be 0·5 mm to give a flow-path ratio of 200:1. This means that in filling the mould, a time can be allowed during which the thickness of the solidified layers on each wall reaches 0·25 mm and the gap is 0·5 mm. Substituting in the formula:

$$t = \frac{0{\cdot}5^3}{0{\cdot}216} = \frac{0{\cdot}125}{0{\cdot}216} = 0{\cdot}575 \text{ s}$$

Thus, for a moulding of 1 mm wall thickness and a flow path of 100 mm, the filling time with material of flow-path ratio 200:1 will be slightly under 0·6 s.

The formula can be generalized using the following symbols:

p_f = flow path in millimetres
p_{fr} = flow path ratio
d_w = moulding minimum wall thickness in millimetres
T_x = heat distortion temperature of the material in °C
T_m = mould temperature in °C
T_c = cylinder temperature in °C
t = filling time in seconds

to

$$t = \frac{(d_w p_f / p_{fr})^3}{8[(T_x - T_m)/(T_c - T_m)]^3}$$

Taking the figures for T_x, T_c, and T_m from Table 4.6, the values for the expression

$$\frac{T_x - T_m}{T_c - T_m}$$

are shown in Table 6.1.

Obviously, the greater the gap through which the mould is being filled, the easier it is to fill. For fairly large areas, the rough approximation mentioned earlier of calculating the filling time on a 50 per cent gap closure can be used successfully and a graphical solution is easiest. Filling time is plotted against

Table 6.1 *Values for $(T_x - T_m)/(T_c - T_m) = V$*

	V	V^3
Low-density polythene	0·3	0·027
Polypropylene	0·32	0·033
Polystyrene	0·35	0·043
Nylon 66	0·52	0·141
Nylon 66 glass filled	0·65	0·275
ABS	0·17	0·004
Acetal	0·32	0·033

the thickness of the two layers of solid material built up on the walls of the mould calculated from the formula:

$$d_s = Vt^{1/3}$$

where d_s = thickness of material on a single wall in millimetres
V = the value from the above table
t = time in seconds.

Using the figures for V from Table 6.1, the results are shown plotted in Fig. 6.1, the vertical coordinate showing the thickness in millimetres of the two layers of the solid material. By ruling on the graph a straight line at one-half the average wall thickness, the point at which it cuts the appropriate curve can be read off and this gives the maximum permissible time for mould filling. Thus, a mould of average wall thickness 2 mm would need to be filled in about 4·5 s using polypropylene.

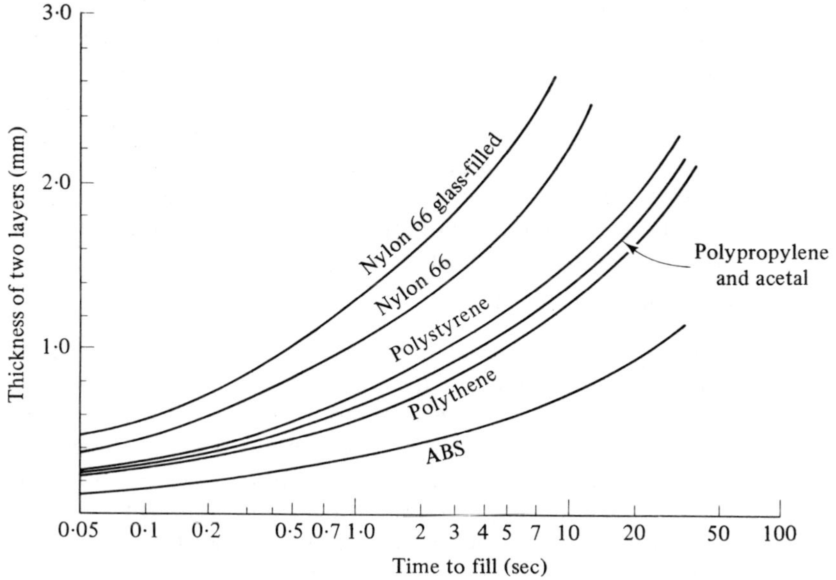

Fig. 6.1 Mould filling: thickness of two layers vs time

Although this method of determining the maximum permissible injection time is a good guide for thin-walled mouldings, it should not be assumed that injection should not be faster than indicated. Except for certain types of moulding, such as ABS components which are to be metal plated and thick-section mouldings in un-filled nylon, much is to be gained by using the most rapid injection rate possible. In this, the limiting factor is often the hydraulic power available in the machine itself. Unless a hydraulic accumulator is fitted to the machine, this depends directly on the horsepower of the electric motor driving the hydraulic pump.

Determination of power requirement

It can be assumed that, for all practical purposes, the fixed-delivery vane pumps used in injection moulding machines operate at about 80 per cent efficiency. The relationship between motor horsepower and volumetric injection rate can then be calculated from the formula:

$$Q = \frac{86\,784 \times \text{hp}}{P}$$

where Q is the volumetric injection rate in cm^3/s

P is the injection pressure in lb/in^2 (on the material, *not* the hydraulic line pressure)

hp is the horsepower of the motor

(Note that the factor 86 784 takes care of the conversion between cm^3/s (c.g.s. units) and pounds per square inch, which are the units most commonly used on the shop floor in the United Kingdom.)

Knowing the time of filling required (from the previous calculation) the volumetric injection rate can be obtained and it then only needs the desired pressure to be known in order to determine what size of pump motor is required, or conversely, whether the machine to be used is suitable for the job. It is essential that the motor should be large enough to maintain the desired pressure for the whole of the injection period. Any falling-off of pressure before the mould is full will lead to the making of faulty mouldings (e.g., poor surface finish near the gate and frozen-in stresses). A method of calculating the pressure required is therefore needed.

Pressure requirements for injection

It has always been taken for granted that a great deal of pressure was required for successful injection moulding and there has been a tendency over the years for the available pressure on machines to be increased from 10 000–15 000 lb/in^2 to as much as 30 000 lb/in^2 in some of the latest models.

A good general rule is to use only as much pressure as is needed to allow injection to take place at the required rate. Having determined the maximum time which can be allowed for injection and thus the quantitative filling rate, the pressure required can be calculated by determining the pressure drop in the feed channels, in the gate, and in the mould cavity itself. Although a complete mathematical breakdown is very complicated, involving, as it does, the pattern of flow behaviour of a viscous fluid at varying flow and shear rates (and thus of varying viscosity) through channels of different dimensions and configurations, a rough approximation is possible by considering the main sections of the channels separately.

The first section to consider is the nozzle of the cylinder and the sprue bush. The channel in the former is converging and in the latter, diverging. From there, the flow may be through the parallel channel of the runners, and eventually the material passes through the narrowest constriction, the gate, into the mould itself. As has been discussed in Chapter 5, there are basically two types of gate, round holes and slits, but for the sake of simplifying the calculation it will be assumed that the round hole is the most common form.

Because round holes produce a lower pressure drop than channels of the same cross-sectional area but of different form, these will be considered for the runners in the calculations and, for the purpose of illustrating the method, will also be considered as being in a straight line, without corners. Obviously, there is always additional pressure drop at corners, and it is for this reason that generously radiused curves and over-size, rounded corners have always been advocated for the runners of injection moulds.

The Newtonian pressure drop in a circular channel is given by the expression:

$$\Delta P = \frac{8\mu QL}{\pi g r^4}$$

where μ is the viscosity of the melt in poise
Q is the volumetric filling rate in cm^3/s
L is the channel length in centimetres
r is the radius of the channel in centimetres
g is the acceleration due to gravity (981 cm/s^2).

For a slit, the pressure drop is given by the expression:

$$\Delta P = \frac{12\mu QL}{gTH^3}$$

where T is the length of the slit in centimetres
H is the thickness of the slit in centimetres
L is the channel length in centimetres

To simplify the above expressions to a form which is more easily used with the units of the moulding shop (*see* above), they can be written as follows:

For round channels or holes:

$$\text{Pressure drop (lb/in}^2\text{)} = \frac{3{\cdot}69\ \mu QL}{100\,000 r^4}$$

and for slits:

$$\text{Pressure drop (lb/in}^2) = \frac{8{\cdot}27\ \mu QL}{100\ 000\ TH^3}$$

In these expressions, μ, Q, L, r, T and H are all in c.g.s. units.

Before the pressure drop can be calculated, it is necessary to know the viscosity of the material as it flows through the channel. From the dimensions Q and r, the rate of shear of the material is first calculated from the expression:

$$\dot{\gamma} = \frac{4Q}{\pi r^3}$$

and by reference to Fig. 4.10 its viscosity at the shear rate can be found. This figure can then be inserted in the equation above and the pressure drop calculated.

Suppose that a mould of 100 cm^3 volume is to be filled in 1 s through feed channels of average radius 0·25 cm and of length 15 cm, using polypropylene:

$$\text{Shear rate} = \frac{4 \times 100}{3{\cdot}14 \times (0{\cdot}25)^3} = 8152\ \text{s}^{-1}$$

From Fig. 4.10, the viscosity of polypropylene at shear rate 8152 s^{-1} is 500 poise.

$$\text{Pressure drop} = \frac{3{\cdot}69 \times 100 \times 15 \times 500}{100\ 000 \times (0{\cdot}25)^4} = 7084\ \text{lb/in}^2\ (496\ \text{kg/cm}^2)$$

Assume a gate of radius 0·1 cm and land length 0·15 cm:
Shear rate in the gate = 127 400 s^{-1}
Viscosity of polypropylene at this shear rate = 40 poise

$$\text{Pressure drop} = \frac{3{\cdot}69 \times 100 \times 0{\cdot}15 \times 40}{100\ 000 \times (0{\cdot}1)^4} = 221\ \text{lb/in}^2\ (15\ \text{kg/cm}^2)$$

Total pressure drop = 7084 + 221 = 7305 lb/in^2 (492 kg/cm^2)

Thus some 7300 lb/in^2 of pressure is used in simply conveying the plastic from the cylinder to the open mould. Something of the order of 1000 lb/in^2 is probably lost by friction in the cylinder, making the total about 8000 lb/in^2.

On average, it is probably fair to say that, of 20 000 lb/in^2 of pressure available at the cylinder, about one-half is used in conveying material to the mould, leaving about 10 000 lb/in^2 for mould filling. Probably one-half of that pressure is lost as the material passes through the various parts of the mould, leaving 5000 lb/in^2 for pressurizing the mould. This pressure will ultimately act against the locking force of the mould and is roughly equivalent to 2–2½ tonnes force per square inch of projected area of the moulding.

From this example, it is now possible to calculate the pump horsepower required to give the desired injection rate (assuming that there is no accumulator on the machine) by substituting the above figures in the expression:

$$\text{hp of pump} = \frac{Q \times P}{86\,784}$$

where $Q = 100\ \text{cm}^3/\text{s}$ and $P = 20\,000\ \text{lb/in}^2\ (1400\ \text{kg/cm}^2)$

$$\text{hp of pump} = \frac{100 \times 20\,000}{86\,784} = 23\ \text{hp}\ (17\ \text{kW})$$

One fact becomes apparent on studying the above calculations, namely, that halving the length of the feed channel (or the land length of the gate) halves the pressure drop in that part of the system. If the radius of the feed channel is altered to the same extent, the change is much greater because the pressure drop is inversely proportional to the fourth power of the radius. Although this affects the viscosity of the melt, the net result is that on doubling the radius the pressure drop is divided by four. This means that in altering gate sizes better control is achieved by reducing the land length than by increasing the radius; and it also indicates that making bigger runners may be more effective in reducing pressure drop than having shorter ones. In the example above, had the feed channels been of radius 0·3 cm instead of 0·25 cm the pressure drop would have been about 5000 lb/in^2 instead of 7000 lb/in^2. However, the weight of the runner system, which is, after all, waste material, would have been 40 per cent greater.

Hydraulic pump

It is important that the hydraulic pump is sufficiently large to maintain the injection pressure throughout the whole of the injection period. If it is too small, although the pressure starts high, it quickly falls to the level which the pump is capable of maintaining. When the mould is nearly full the rate of injection slows, but as the pressure builds up, a small quantity of molten plastic is forced in, near the gate. This is unlikely to fuse together properly with the material already there and the well-known fault of splay marking results. The more viscous the melt at this time, the more difficult it becomes for pressures to even themselves out and so areas of stress develop near to the gate. These can be detected by accurate measurements of the density of the material in this part of the moulding. If the stresses are at all high, breakdown under service conditions, especially in the presence of an environmental stress-cracking agent, can occur.

The same effect, incidentally, occurs if the check-ring or back-flow stop valve on the end of the screw is faulty. The operator will first notice that the screw 'bottoms'—that is, it moves forward to the limit of its stroke, but produces a short moulding. He then increases the feed to compensate and is able to fill the mould under a low back pressure. When the pressure in the system reaches the point where leakage occurs at the check ring it stays constant until enough material has passed backwards along the screw to block further leakage, and then increases. A little more material is then injected but it is at a higher viscosity; it does not distribute itself evenly and causes an area of stress, exactly as if the pump were too small.

With adequate pump capacity, the mould will fill in the time allotted, the pressure remains constant, and flow through the gate stops. The gate immediately freezes and then material can neither enter nor leave the mould. Screw-back starts immediately and the moulding produced is free from blemishes and frozen-in stress. In addition, a saving in cycle time may be effected because the plunger dwell time is reduced. Mouldings will behave much better in service and will withstand the effects of environmental stress-cracking agents much better.

Trimming the process

Having arrived at the correct requirements of pressure, time, and rate for successful moulding, it is necessary to examine their interaction and to make any final adjustments; in other words, to trim the process. Efficiency in injection moulding is dependent upon attention to fine detail as much as, or even more than, on the availability of sophisticated control equipment or other elaborate machine refinements. For example, the mould locking force should be carefully adjusted to a level of no more than is necessary to hold the mould closed during injection, and the injection pressure should be no more than that required to give the correct rate of injection. Yet, how often, in practice are these parts of the process carefully and thoughtfully operated?

Most material manufacturers will provide data sheets for their products in which recommendations are made for the conditions under which best results are obtained. Such conditions should always be thought of as 'starting conditions'. No two machines are exactly alike so far as temperature regulation, heat transfer, and pressure application are concerned and the operator or setter must be prepared to make an experimental approach to optimizing conditions. Neither can a mould be taken from one machine and run on another one at exactly the same process control settings. Trials are always necessary to obtain the best results and the ultimate process conditions may deviate considerably from those found correct on another machine.

The same observations can be made in respect of interrupted production runs where a mould is taken off a machine but is subsequently refitted to the same machine to continue the production programme. Even if all the previous control settings have been carefully observed and noted, there is nearly always a slight variation with the later run. The reasons for this are many and intricate and could include such minor items as hydraulic oil temperature, the accurate setting of mould locking, cylinder temperature control, and injection rate setting. Looking once more at the way the machine functions will help to explain some of the faults that can occur.

Balancing process variables

Having established that accurate control over the whole moulding process is necessary for efficient moulding and that this accuracy has to be maintained for the duration of the production run, what are the factors that can cause variations? These are inherent in the materials being worked with and must be taken into account when balancing the process variables.

(a) *Hydraulic oil.* Starting with the hydraulic oil, this is slightly compressible and its viscosity changes with temperature. If there is a deviation from normal working temperature and pressure, the various control valves will be affected and the rate of flow of oil will change. Because of its compressibility, doubling the pressure will not of necessity double the velocity of the oil flow and, as it is sheared in the pressure control systems, it becomes hotter. Thus, slight variations in the temperature of the oil and in the operation of the pump (for example, a variation in the voltage of the electrical supply) can have an enhanced effect on the rate of flow of the hydraulic fluid.

(b) *Mould locking.* This change can affect both ends of the machine—the mould locking arrangements and the mould filling. A lowering of the viscosity of the oil, accompanied by a rise in its temperature, causes it to flow more rapidly and so speeds up the mould closing and increases the effective locking force. If the viscosity of the oil increases, the reverse happens and the machine becomes sluggish. It is even possible for the amount and speed of mould locking to affect the rate of mould filling because this often depends on the speed at which the air trapped in the mould can escape. Extra heavy mould locking can block the natural escape of air at the mating surfaces and so prevent the entry of material. Conversely, extra-light locking can help to increase filling rate by making the escape of air from the mould easier. Variation in the amount of mould locking can also cause a variation in the size of the moulding because, although the steel of the mould may be hardened, it is very slightly elastic and compressible and under the heavy pressure of locking the mould cavity could become smaller. Very light locking will possibly allow the mould to open slightly during injection and thus give a larger or heavier moulding.

(c) *Injection.* At the injection end of the machine, the same changes in the oil can cause faster or slower movement of the screw/plunger system and thus affect the injection rate. This has an enhanced effect because of the pseudoplastic nature of the plastic melt. A slightly increased rate of injection brought about by more rapid flow of oil through the system increases the rate of shear of the material and thus decreases its viscosity. Its flow into the mould is changed and this may be sufficient to alter the positions of flow lines and weld lines in the moulding or even to cause in-bowing or out-bowing of the walls of a moulded box.

A change in oil temperature also affects the rate of flow through flow control valves. No flow control valve is fully temperature-compensated and two effects are noted. Because the flow control valve is applying shear to the oil, heat is developed and the body of the valve gets warm. The oil also becomes heated and the total result is shown as a quite considerable increase in injection rate.

In the moulding of a pocket magnifier from acrylic material in a single impression mould, a slight blemish, identified as a weld mark, was observed at the side of the lens (Fig. 6.2). This was caused by the joining of two flow paths of material, one following the bezel of the lens and the other proceeding across its face. By altering the flow control valve to reduce the rate of injection, the blemish could be moved to a position where it became less objectionable, in the angle between bezel and lens. As moulding proceeded, the injection flow control valve became warm and the blemish moved back to its original position. By adjusting the valve again it was possible to move the blemish and by periodic adjustments it was possible to produce acceptable mouldings.

(d) *Making adjustments*. Adjustments such as those described above may be necessary when critical mouldings are being produced. This becomes tedious if manual operation is necessary and an automatic method is desirable. A closed loop system that monitors the moulding pressure and makes the appropriate adjustments will certainly assist in the making of blemish-free mouldings in cases such as this, but it is expensive. Perhaps if the hydraulic valve designers can produce control valves which are self-compensating with respect to temperature, some of the difficulties could be overcome. However, many of the problems of moulding, of which the above account is typical, are brought about because the operator attempts to produce perfect mouldings before the machine has had a chance to settle

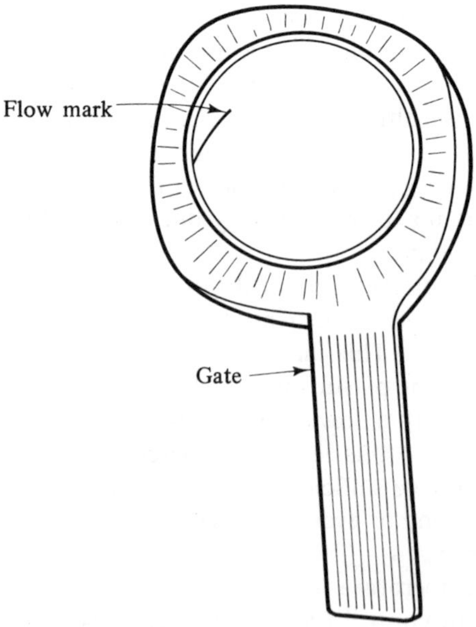

Fig. 6.2 Pocket magnifier showing position of flow mark

down. On start-up, unless an oil heater is fitted in the tank, the oil may be cold. Valves and pipe-lines are also cold. Without waiting for the oil to warm up, full cooling of all the machine parts is turned on and it may then take several hours to reach equilibrium, let alone optimum working temperature. During this period minor variations occur, not only in the quality of the mouldings but also in their weight. Every effort should be made to ensure that the required machine conditions are obtained as soon as possible and, for this, thermostatically controlled valves on the coolant supply to the oil coolers and thermostatted oil heaters may be more effective and considerably cheaper in achieving good quality mouldings than closed loop, computer-controlled systems.

Switching systems and process variables

The other items in the control of the machine that cause variation are switching systems and mould locking. Of the former, sequence and time control are more susceptible to variation than is temperature control. This is generally very effective and it is only infrequently that a heater band or a thermocouple ceases to function. However, it has been known for a thermocouple to be inadvertently pulled out of its pocket, causing a rapid rise in temperature of the section of the cylinder which it controls. Care also needs to be exercised if two or more heater bands are controlled by the same thermocouple, because this can produce a tendency for a hot spot to develop in the part of the cylinder that is further away from the thermocouple. In the same way, a compound heater band, that is, a heater band with two or three sections connected in parallel, can cause a hot spot to form if one of the sections, particularly that nearest to the thermocouple, breaks down.

Microswitches, relays and solenoids have already been discussed, but it should be emphasized again that dust and dirt in any of them can cause sluggishness in the action of the machine. Noting this condition, the operator will increase the pressures all round in an attempt to speed up the cycle, with an unfortunate effect. All the carefully calculated and set conditions are overridden and a high degree of variation results. The effects of dust and dirt are, of course, cumulative and may well account for the fact that a mould taken off a machine and then put back on a month or two later, will rarely work at the previously set conditions.

All mechanical switches, whether timers, microswitches or relays, are susceptible to fatigue in their contact springs. There is also the danger of dirty or spark-pitted contacts developing, and these reduce the voltage transmitted and cause the relays and solenoids to move more slowly. Again, the operator compensates by increasing the system pressure with the result described earlier.

Mechanical systems and process variables

The other important cause of variation in the moulding process is a purely mechanical one: the friction developed in the actual movement of carriage, mould platens, or moving parts of the mould.

(a) *Carriage movement.* Only the minimum of force required for proper operation should be developed in the hydraulic cylinders that move the carriage and the mould platens. Too great a pressure exerted on the sprue bush by the nozzle can cause damage to either part and is particularly prone to damage extended nozzles. Too rapid a rate of forward movement of the carriage can cause hobbing of one part into the other. Both conditions put undue strain on the carriage bearings, causing the slide and bushes to wear. The stage is sometimes reached where, when the nozzle contacts the sprue bush, the pressure forces the two out of line and material escapes at the joint. To compensate for this, the operator increases the feed setting, but because the same amount of material does not escape at each shot, there is a variation in weight of the mouldings from shot to shot.

(b) *Mould mating.* Incorrect mating of the mould surfaces is another common cause of variation in the process. Badly aligned platens do not slide easily on the tie bars and a great deal of the available mould closing force is used up in moving them. This is frequently caused by incorrect bedding of the machine on the floor of the moulding shop, or some unevenness preventing the machine from sitting perfectly level. An accumulation of dirt and dust on the tie bars may also be a contributing factor in reducing the available locking force. The same effect can occur in the mould itself. Guide pins may not slide easily in their bushes because of bad alignment of the mould parts or because of differential temperatures on the two halves, and the mould may not close properly. If this occurs, the mould may flash, but even if it does not flash, the moulding will be thicker than it should be. The effect is really the same as if there were not sufficient oil pressure for mould locking, as already discussed.

(c) *Tie-bar alignment.* Occasionally, the machine tie bars or mould dowel pins become sufficiently out of line to cause a thickness variation from side to side of a moulding. This kind of fault should not normally occur if the machine has been properly built and the mould accurately made and set, but it may become evident if a tie bar has to be replaced. It can also occur if the platens have become distorted because of bad mould setting or if there is undue wear on the tie bars or the platen bushes. When re-setting the machine after replacement of a tie bar, it is most important that the tension on all the tie bars is set at the same level and this is best achieved by the use of strain gauges. Unless the tension is the same in all the tie bars there is a great danger of fracturing one or more of them, because at full lock they may be stressed beyond their elastic recovery limit.

Feed setting

Another variable in the process that needs consideration is the amount of feed: the volume of material prepared for the next shot and stored at the front end of the cylinder. Some moulders advocate keeping a cushion of material at the front of the cylinder, against which the screw can act during the holding time or plunger dwell time, with the intention of allowing for the reduction in volume of the material in the mould as it cools. This may be desirable with some thicker section mouldings using a low viscosity melt, but in general, it is not good practice. The cylinder should be emptied at each shot and the feed adjusted to give the exact shot weight. In setting the machine, when the correct pressure settings have been made and injection is taking place at the correct speed, the feed should be adjusted to give the full shot weight when the screw 'bottoms'. This is particularly important when moulding heat sensitive materials such as PVC and acetals, but it is best to follow the procedure for all materials and jobs, except those mentioned above. Waste is reduced and nozzle dribble and tailing (formation of a thin filament of material from the nozzle when the mould opens) become less of a nuisance.

Interdependence of process variables

From this consideration of the fine trimming of the process and the balancing of process variables, it will be seen how dependent each of these is on the others. The bigger the machine, the longer the time that it takes to react and, therefore, the gentler should be the adjustments made to the control system. This is nowhere more important than in the control of cylinder temperature. At their moulding temperatures the viscosities of the materials react quite sharply to small changes of temperature, and when this effect is amplified by using high shear rates it will be appreciated how important temperature control can be. It can, however, be said with some justification that the *actual* temperature indicated is not of primary importance, although to have it indicated helps in setting. However, it is important that, having found correct conditions *as indicated on the machine*, those conditions can be maintained. Ultimately, although guidance can be obtained from manufacturers' literature and from calculations, the success of injection moulding depends on the ability of the operator to recognize the conditions under which correct quality mouldings are being obtained and to maintain those conditions. Keen observation is a necessary quality in a good operator and, in noting down the conditions under which the machine is operating successfully, nothing should be overlooked. The more obvious conditions will not, of course, be forgotten, but such matters as the temperature of the oil and the cooling water, perhaps even the voltage of the electrical power supply, may be significant. In fact, the state of the weather—hot, cold, wet, or dry—could influence the obtaining and *maintaining* of optimum conditions.

7

Injection moulding of thermosetting materials

Although thermosetting moulding materials have been available for nearly 70 years, it is only recently that any great success has been achieved in moulding them by injection moulding techniques. Compression moulding has been the standard method and even though it can be made fully automatic, the time cycle has tended to be long. Transfer moulding, which can be thought of as 'one-shot injection moulding' has been widely practised, particularly for the moulding of components with integral holes or moulded-in inserts, but, even though the time cycle is considerably shorter, the amount of waste material tends to be excessive. Injection moulding appears to offer a solution to both these problems.

In the years prior to 1940, some progress was made in the injection moulding of thermosetting materials, chiefly those based on phenol/formaldehyde resins, using the rather rudimentary plunger machines then available. It was the advent of the screw/plunger machine in the period 1955 to 1965 that began to see more rapid advances in the technology. Today, the injection moulding of thermosetting plastics is rapidly overtaking in importance the slower, more wasteful, conventional compression and transfer moulding.

Problems with thermosetting materials

Thermosetting materials pose some problems for the injection moulding technologist. In the initial plasticizing stage of moulding, the materials act in a similar way to thermoplastics. They become viscous fluids which, under pressure can be made to flow quite readily, and the melts have similar pseudoplastic properties to the thermoplastics. Their minimum viscosities at shear rates of about 1000 s^{-1} are in the region of 10 000 poise (somewhat higher than the easier-moulding grades of thermoplastics), and this brings them near to the so-called 'treacly' range of viscosity. Being thermosetting materials, however, as soon as the temperature begins to approach their moulding temperature (135–140°C for aminoplasts and 160°C for phenoplasts) the chemical change of thermosetting commences and the viscosity rapidly increases until the plastic becomes completely solid. At a temperature of 100°C this may take 10–30 min, but at 160°C it is very rapid and may take less than 1 min.

Because of this chemical change and the accompanying change in viscosity, any calculations based on rheological properties become extremely complicated, if not virtually impossible, and at the present time the technology of the injection moulding of thermosetting materials is largely a matter of trial and error, coupled with the need for extremely keen observation.

Machines for thermosetting plastics

Some modifications are necessary to the standard injection moulding machine to make it suitable for the moulding of thermosetting plastics. Instead of aiming at complete plasticization of the material in the cylinder, the objective is to raise the temperature of the cold moulding powder fed to the hopper to such a level in its passage along the cylinder that, when subjected to the high pressure of injection, it will become fluid as it passes through the flow channels into the mould cavity. During its passage through the nozzle of the cylinder, frictional heat is developed, and this gives the initial fluidity to enable the flow to continue through the sprue bush and runners. Exactly as in the moulding of thermoplastics, the screw is used first for mixing and heating of the material and then as a plunger for applying the high pressure needed for injection. The pressure is held after the mould is full until cure is complete and then screw-back occurs to prepare the next charge.

Instead of the conventional electrical heater bands, the cylinder is provided with a jacket through which water or oil can be circulated at a controlled temperature. Like the electrical heating, this type of temperature control is zoned, so that the feed end can be at a different temperature from the centre of the cylinder and from the nozzle zone. Typical temperatures for urea-formaldehyde moulding powders would be 75°C at the feed hopper, increasing to 95–100°C at the nozzle. Accurate control of the cylinder temperature is of the greatest importance because of the susceptibility of the material to cure prematurely if too high, and undue lengthening of the curing cycle if it is too low. Another complication to be taken into account is the variation in the effect of mechanical heating that can occur with different cylinder temperatures.

The machine must be capable of rapid injection at high pressures of the order of 25 000 lb/in^2 (1750 kg/cm^2). This must be maintained constant over the whole period of injection (generally necessitating the use of a hydraulic accumulator) and must then be capable of being reduced for the holding period. The machine must be capable of rapid sprue break because, during injection and cure, the nozzle is kept in contact with the hot mould and this transfers heat back into the nozzle, causing pre-cure of the material resident there. Sprue-break allows a period when the nozzle can cool off after the injection and cure time, and when the machine is running smoothly with effective sprue break it is found that the nozzle will stay at a fairly steady temperature.

Unlike in thermoplastic moulding, the mould in the injection moulding of thermosetting plastics is kept at a high temperature. It does not have a cooling effect, but a heating effect on the plastic melt. Mouldings can be removed hot from the mould as soon as curing is far enough advanced for the components to keep to their required shapes. Optimum temperatures have to be determined by a process of trial and error, but in general they are very similar to those recommended by the materials manufacturers for the compression moulding of their products, namely 140°C for urea-based materials, 150°C for melamine, and 160°C for phenolics.

Thermosetting moulding materials

Thermosetting moulding materials consist of thermosetting resins mixed with fillers, lubricants, pigments, etc., and small quantities of catalysts to bring about the chemical change of cross-linking which takes place on heating and which gives this range of plastics their particular properties. The cured or cross-linked resin is insoluble and infusible and cannot be re-ground, like thermoplastics, and used again for moulding. The fillers are included because, in general, the resins are not strong enough to be used alone and also to make the product cheaper.

Quite a large number of resins are used, but by far the most common are those based on phenol and formaldehyde (often called PF resins) and urea and formaldehyde (referred to as UF resins). Closely related to the urea resins are melamine/formaldehyde materials which, with urea plastics, together form the group often called aminoplasts. The term phenoplasts is sometimes applied to the resinous products of phenol and formaldehyde. Alkyd resins, based on diallyl phthalate, or on similar polyesters, are also used as the basis for thermosetting moulding materials.

Characterizing thermosetting moulding materials. The behaviour during moulding of thermosetting moulding materials is characterized in two main ways: cure time and time of flow. The cure time, determined by moulding a standard item in a standard mould at set conditions, is the time taken from the moment of complete closing of a compression mould containing a charge of the material to the time when cure is complete. The test for complete cure is that the moulding is stable, does not blister on removal from the mould, and yet shows no signs of being overheated. Three general grades of moulding powder are produced being classed as fast, medium, and slow cure.

The time of flow is determined in a standard beaker mould and is known as the cup flow time. The required amount of the moulding powder is placed in the standard cup flow mould which is maintained at a set temperature. The mould is closed using a set pressure, and the time taken from when the full pressure is first applied to the time when flow of material in the mould ends (this is indicated by means of a micrometer gauge on the mould platens). A free-flowing material (or soft flow) may take about 10 s, a medium flow 15 s, and a stiff flow about 20 s.

Thermosetting moulding materials are also characterized by their fillers. General-purpose phenolics normally contain about 50 per cent of wood flour and high-impact materials may contain fibrous organic fillers. Electrical grades often contain glass fibre and heat-resistant grades have asbestos or other mineral fillers. Most high-quality aminoplasts are filled with comminuted pure alpha cellulose, but slightly inferior materials from the point of view of colour contain wood flour. Alkyd materials are normally filled with glass fibres of different fibre length, according to the particular properties required. Most of these types are specially modified to make them suitable for injection moulding.

It is possible to mould nearly all types of thermosetting moulding powders by injection techniques, but the best results are obtained with materials of soft to medium flow and medium cure. Fast curing powders tend to set up before the mould is filled, while slow curing powders take too long to mould. The modifications to materials that have enabled satisfactory injection mouldings to be made, have resulted in materials of the correct combination of properties to give high outputs at the lowest possible temperature. Cycle times of around 20 s are possible with moulding sections $\frac{1}{3}$ in (7·5 mm) thick and these would take three or four times as long to cure, even using radio-frequency preheating, if moulded by ordinary compression techniques.

Process details

To prevent premature heating of the material near the feed pocket of the machine (or even in the feed hopper), it is wise to cool the rear portion of the cylinder. The remainder of the cylinder is carefully maintained at a temperature of between 75–80°C and 120°C, according to the type of material. For aminoplasts, the lower temperatures are used and the cylinder can be water-heated, but oil heating is necessary for heating the cylinder for phenolic materials. With zoned temperature control it is possible to arrange for the best possible temperature gradient to be maintained along the cylinder and for the net effect at the nozzle to be one of cooling rather than heating.

(a) *Temperature effects.* Control of the cylinder temperature is most important and cannot be too highly stressed. Too little heating allows unplasticized material to collect in the screw flights where it becomes compacted. In these conditions, frictional heating by screw rotation can be great enough to cause pre-cure in the cylinder. If the powder is plasticized but is not quite hot enough to enable it to flow freely, it may precure in the nozzle and mould feed channels due to extra frictional heating during injection and thus may prevent complete mould filling.

On starting up, it is a good general rule to set the cylinder temperatures on the low side and gradually to raise them until signs of overheating appear. This may show as slight gassing of the material when purging the cylinder or as a gradually increasing injection time, indicating that there is some stiffening of the melt. An indication that there may be some pre-cured material at the front of the cylinder is given on screw back if the screw rotates slowly for the first few revolutions and then speeds up. On the appearance of any signs of pre-cure it is wisest to purge the cylinder immediately. If this is not done, the powder may set up solid and become extremely difficult to remove. The cylinder temperature should also be reduced again until optimum conditions are obtained.

(b) *Effect of the screw.* The screw speed is important because, with the preheated material becoming very viscous, a great deal of mechanical heat is developed, causing pre-cure. Best results are obtained if screw back ends as cure time ends so that the plasticized material does not have to remain for a long time in the cylinder before being injected.

For small machines and short curing times, screw speeds of 70 to 100 r.p.m. can be used, but with large machines, less than 50 r.p.m. should be used so that undue frictional heating does not take place. If the whole of the cylinder capacity is not being utilized for the required shot weight, with some types of machine and certain grades of material it is often found best to have a temperature difference of about 20°C along the cylinder, and this will compensate for any variations in frictional heating that may occur.

(c) *Rate of injection.* Ideally, injection should be as rapid as possible, but running at maximum rates presents problems. First, the material must be at the correct viscosity or else there will be too much frictional heating in the mould feed channels and pre-cure will result. Second, the higher the injection rate, the greater is the pressure needed, and this has meant the development of machines capable of injection pressures of 25 000 lb/in^2 (1750 kg/cm^2) or more. Thirdly, mould venting has to be extremely good. Given rapid injection, this means that less work and plasticization is required in the cylinder, the necessary heating being given by friction in the feed channels. Material enters the mould at the temperature required for curing, better cavity filling occurs, and cure time is reduced. Holding time can be reduced or dispensed with and frozen-in strain and excessive mould packing are avoided. Weld lines and gassing may also be reduced.

(d) *Mould temperature.* A short moulding cycle is favoured by a high mould temperature, but if too rapid curing takes place the mouldings are liable to exhibit porosity in the thicker sections and some burning may take place. Flow marks, warping, and excessive shrinkage may occur with too rapid curing.

As with finding the optimum cylinder temperature, so it is advisable to start with the mould at a somewhat lower temperature than would be used in compression moulding and to increase it gradually until best results are obtained. Suitable starting temperatures would be 145°C for urea materials, 150°C for melamine, and 155°C for phenolics. Still raising the temperature will result in the first signs of pre-cure appearing, and when this happens the best procedure is to make minor adjustments to injection speed and pressure before reducing the mould temperature. Sometimes it is possible to find slightly better conditions in this way.

(e) *Other processing factors.* A certain amount of back pressure on the screw during screw back can be helpful in increasing the amount of mechanical heating developed. This must be very carefully regulated and harmonized with the screw speed so that the best latitude possible in moulding conditions can be obtained. The time during which material resides in the cylinder is also important. The smaller the shot in relation to the plasticizing capacity of the machine, the longer will be the residence time and the greater the possibility of pre-cure. However, if the cylinder conditions are properly regulated, the material will be plasticized to its correct degree only as it reaches the front of the cylinder and, because it will all be used for the shot for which it is prepared, it will not be pre-cured.

Mould design for thermosetting moulding

Care must be taken in mould design to eliminate all factors which could contribute to the production of faulty mouldings. Many of the principles that apply to thermoplastic moulding also apply here, with the exception of mould cooling. On the other hand, mould heating must be just as carefully designed as mould cooling so that there is no possibility of uneven temperature distribution. Hot spots can be a cause of pre-cure.

(a) *Sprues and runners.* Sprue and runner systems should be as short as possible to reduce wastage (unlike thermoplastic sprues and runners, these cannot be recycled). Full round runners are preferred as they reduce frictional losses during filling and give a smaller surface area/volume ratio, thus reducing surface heat absorption. As in all moulding, a compromise must be made between easiest filling, longer curing (runners cannot be removed from the mould until they are cured) and waste. Aminoplasts require somewhat larger sprues and runners than phenolics.

(b) *Gates.* The gate is just as important as in thermoplastic moulding. Here again, compromise is necessary. A smaller gate leaves a less obtrusive scar on a moulding, but produces a greater pressure drop during filling and more frictional heating. It also tends to get more abrasion, thermosetting materials being generally more abrasive than thermoplastics. As a general rule, the gate should be as small as possible, consistent with obtaining good mouldings, and obviously the land length should be as short as the configuration and materials of construction of the mould will allow. Provided that gating is into the thickest section of the mould, its position is relatively unimportant.

Flash gates and fan gates can assist in the distribution of the material in producing mouldings having flat areas. Core pins or acute angles in the direct flow path of material entering the mould can increase resistance to mould filling and should be taken into account when positioning the gates of difficult shapes.

(c) *Multi-impression moulds.* In multi-impression moulds, the flow paths should all be of the same length and balanced so that all cavities fill simultaneously. Flashing is liable to occur where flow paths are out of balance.

(d) *Mould venting.* The most important aspect of mould design, as with thermoplastics, is mould venting. A mould cannot be filled quickly unless the air in it can escape with equal facility. Being of higher viscosity, the melt will not escape as easily as most thermoplastics from small gaps, and vents can be considerably larger. A typical vent would be of thickness 0·0015–0·005 in (0·05–0·15 mm) and length up to 0·75 in (19 mm). The vent should be relieved on the down-stream side to prevent it becoming blocked. Venting can also take place along ejector pins and at the normal parting lines of the mould, and if any slight witness marks occur they can nearly always be easily removed. 'Breathing' the mould, that is, opening it slightly and quickly reclosing, as used in compression moulding to allow gas to escape, can also be used with great care in injection moulding of thermosetting materials.

Faults in thermosetting mouldings

Most of the reasons why faulty mouldings occur have been discussed either in the preceding paragraphs or elsewhere in the book. If variations take place which can be attributed to the functioning of the machine, Table 8.1 will be found helpful. There are, however, some faults that are quite specific to thermosetting materials and they will now be briefly described and commented upon.

(a) *Blistering.* This fault, in which sections of the moulding form into a double skin resembling a blister on the skin, is the result of volatile material within the moulding expanding when the pressure of the mould is reduced before the skin of plastic is strong enough to withstand the internal pressure. When thermosetting materials cure, there is nearly always the liberation of volatile matter, such as water. There may also be other decomposition products present and these are collectively referred to as 'gas'.

If the moulding is not fully cured when the mould is opened and 'gas' is present, blistering will result. All the causes of undercure and of decomposition previously discussed should therefore be considered and the appropriate action taken. Generally, a few seconds longer cure will be sufficient to reduce the fault, but if this does not succeed it will be necessary to check the mould venting. 'Breathing' the mould can also be used to alleviate this fault, but must be carried out very carefully.

(b) *Gas marks.* These are marks similar to burn marks in thermoplastic mouldings that occur sometimes when mould venting is inadequate. Stiffer flow materials tend to produce less gas than the freer-flowing ones and will sometimes reduce the trouble. Slower injection allows more time for gas to escape and reduces the effect. Running at the minimum possible temperature reduces the tendency of the material to decompose and reduces the amount of gassing. However, all these possible cures can have their side effects and need to be applied with great care.

(c) *Mould staining.* This fault, like gas marks, is associated with the excessive generation of gas caused by slight decomposition of the material. The general method of reducing the effect is the same as for gas marks. Some additives also tend to cause staining, in which case frequent cleaning of the mould is required.

(d) *Short mouldings.* As with thermoplastic moulding, short mouldings are the result of difficulty in mould filling. The same remedies apply—reducing any back pressure in the feed system which could prevent adequate pressure being applied to the mould, and making sure that either air trapped in the mould or material setting up prematurely in the mould channels is not preventing proper filling. It must be remembered, too, that the thermosetting material might be pre-curing because of the generation of too much heat. If the correct conditions have been established and short mouldings suddenly appear, it may be that the non-return valve on the screw has become inoperative because of some foreign matter under the seating. The material should also be checked to make sure that it is not of too stiff a 'flow'.

Other faults such as excessive flash, flow marks, distortion, poor surface finish, etc., are caused by much the same conditions as apply to thermoplastic mouldings. Faults such as sink marks and voids do not normally occur with thermosetting materials, because there is not the same change in volume on processing found with thermoplastics. However, if thermosetting components have been badly moulded, with excessive residual stresses, then the same effects of breakdown in service, particularly in aggressive environments, will occur as with thermoplastics. A great many problems arising because of service failures can be traced to poor moulding techniques.

8

Causes, diagnosis and cure of faults in injection mouldings

The major faults in injection mouldings have already been described in the glossary in Chapter 1. The descriptions were, of necessity, brief and in this chapter they will be enlarged and detailed and their causes related to material, mould or machine. In addition, numerous problems that occur in the day-to-day working of a moulding shop will be considered. Typical of these are difficulties in the ejection of mouldings, sticking in the sprue, and incorrect plasticization of material. While they may not cause actual faulty, or at least unusable, mouldings, these problems sometimes take a long time to diagnose and rectify, thus affecting the efficiency of the plant. A check-list will be included to cover the many features of machine, material and mould that can contribute to what is one of the major problems of injection moulding, *viz*. the variation from shot to shot and the variation over a period of time in the weight and dimensions of components.

Burn marks

If an area of a moulding becomes oxidized, discoloured, or blackened, the first thing to do is to examine the venting of the mould. Air trapped in the mould is heated to a very high temperature as it becomes compressed by the material entering the mould. It reacts with the hot plastic and brings about oxidation. There may not be sufficient air to cause complete combustion of the plastic to carbon dioxide and water, and the hydrogen component of the material reacts first, leaving a black deposit of carbon. If the burn marking is allowed to go unchecked it will eventually cause oxidation or burning of the metal of the mould and the surface will become pitted. Even if the burning ceases because the air finds a way of escape, the surface finish of the mould may be blemished and the part will need to be repolished. Figure 8.1 shows a typical example.

Air trapped in a mould may be the result of poor mould design leaving a zone of the cavity not swept by the incoming wave of plastic. More frequently,

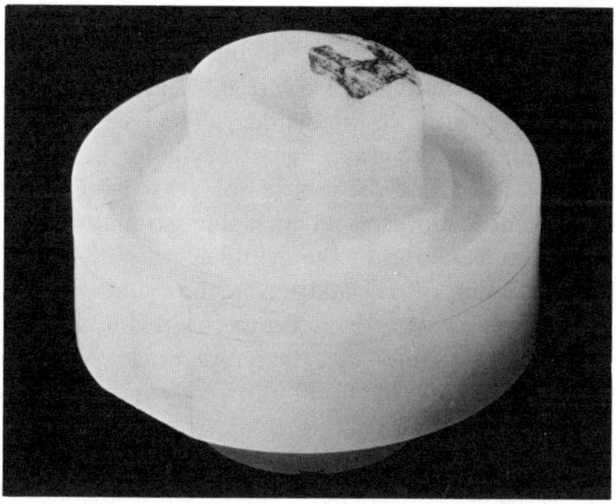

Fig. 8.1 Burn mark caused by poor venting

it occurs because the mould maker seems to do his job too well. Normally, air escapes at the mould mating surfaces and along ejector pins or stripper plates. If the latter are too well fitting or if the former have been too accurately polished this means of escape is barred. (It is interesting to note that the mating surfaces of moulds are generally the easiest surfaces to polish and therefore get the most attention). A simple expedient, sometimes resorted to by a mould maker (and occasionally recommended by the mould designer), is to arrange for a gap of 0·0005 in (0·0125 mm) about 1 in in length (25 mm) to be made at a suitable place. If this were the only vent for a mould of capacity 100 cm^3 filled in 1 s the velocity of the escaping air would be about Mach 1 (1000 f/s or 330 m/s). Obviously, great heat would be developed at such a vent and for this rate of filling, which is not very great, much wider and thicker gaps are needed.

Most plastics will escape very little, if at all, through gaps of 0·0015 in (0·0375 mm) and the time-worn expedient of the cigarette paper between the mating surfaces has much to commend it. It is nearly always desirable that the mating surfaces should be left unpolished, possibly with fairly coarse tool marks, so that natural venting can occur.

Another cause of burn marking is the trapping of air in the nozzle of the cylinder when suck-back or decompression is used incorrectly. This is the slight backward movement given to the screw after screw-back, which helps to reduce nozzle dribble. If the suck-back is too great, the material in the nozzle may be oxidized and will show as a burn mark in the moulding. Its appearance will be different, as if it were 'under the skin' of the moulding and its position will vary slightly from shot to shot. Normally, it will not have any soot associated with it.

The third type of burn marking occurs because of oxidation or decomposition of the material within the cylinder and is more common with PVC, acetals, and nylon than with other materials. With PVC it appears as a dark streak extending through the moulding, varying in position from shot to shot. The cure is to reduce the amount of electrical heating and to increase the amount of mechanical heating in the cylinder. The best way to do this is to work with a reverse temperature gradient in the cylinder, keeping the rear zone hot, the centre zone cooler, and the nozzle hot. It is also important to make sure that the screw almost 'bottoms' at each shot, so that no material resides for too long in the front section of the cylinder.

With acetal the first sign of degradation in the cylinder is a darkening of colour from white to yellow or brown. Temperatures should be reduced and the cylinder purged. At the same time, the temperature controllers should be checked to ensure that they are functioning correctly and that heater bands are working properly.

Finally, shear heating in the gate: this is often blamed for burn marks or signs of decomposition around the point of entry into the mould cavity, but these are much more likely to be the result of overheating in the cylinder. Experiments and theoretical work have shown that shear heating giving a temperature rise of up to 20°C can occur in long capillaries at high shear rates. In injection moulding, the capillary, in this case, the gate, is very short and although the shear rate is high, there is probably no net increase in the material temperature—the mould filling is adiabatic. The temperature may rise on the compression side of the gate because of the high pressure being applied, but immediately after passing through the gate the material is decompressed, a process requiring the application of heat.

Cracking or crazing

This fault occurs only when there is residual stress in a moulding which may be affected by an aggressive environment. The term 'aggressive' may be considered a misnomer for such environments as a low concentration of ozone in the atmosphere or for a seemingly bland material such as Vaseline petroleum jelly. However, both are aggressive to stressed plastics. The only way to avoid cracking and crazing is to produce stress-free mouldings and this has been discussed in an earlier chapter.

Delamination

Like cracking and crazing, this fault is a sign of a stressed moulding in an aggressive environment. It is sometimes seen in domestic hollow-ware which has been subject to household detergents. For example, in using a washing-up bowl, the liquid detergent is nearly always poured into the centre of the bowl before water is added. The high concentration of detergent and the sudden rise in temperature as the bowl is filled with hot water, coupled with the frozen-in stress in the base of the bowl cause the surface to peel like the skins of an onion and to show the signs of delamination.

Another common cause of delamination is the excessive use of mould release agents, particularly those based on silicone oils and resins. These should only be used very sparingly, not at all in the moulding of polythene, and if sprue-sticking or other extraction difficulties occur, a check should be made of moulding technique. With a well-designed mould and correct moulding techniques, mould release agents should not be required.

Contamination of one material by another (e.g., polythene and polytyrene) often manifests itself as delamination.

Distortion

By distortion in this context is meant the taking up by a moulding of a shape other than that required. It is different from warping, in that it implies that the shape never was correct. The commonest cause of distortion is that the moulding has not been allowed to cool sufficiently before removal from the mould and that it has been physically pushed out of shape by ejectors or other means. For example, ejector pins acting on a still-soft moulding will stretch the material, or a very soft portion of a moulding prematurely taken from the mould may flow slightly. Although jig cooling, either in the mould or in special jigs after extraction of the moulding, would seldom be recommended, it is important to make sure that the cooling period is long enough to allow the moulding to become strong enough to handle.

Flash

The escape of a thin web of material at parting lines, around ejector pins or at other parts of the mould, can be caused in several ways. It could be the material, the machine, the mould, or the process that is at fault. Taking these in order:

(a) *Material at fault.* Manufacturers are very careful about the quality of their products but very occasionally it is possible for a quantity of material of the wrong melt flow index to be included in a consignment. If flashing occurs shortly after recharging the feed hopper and if no other changes have been made, the material can be suspected. Too easy a flow or too high a temperature are common causes of flashing. If re-work has been used, the quantity should be checked because passage through the machine can raise the melt flow index of the material. The fault may also occur when changing colour for, as previously discussed, some colours seem to take up heat in the cylinder more quickly than others.

(b) *Machine at fault.* If flashing occurs, the first thing to check is the mould locking. A change of temperature with a toggle locking machine might be sufficient to reduce the mould locking force, but it may also be that the mould parts are binding on their guide pins and preventing them from coming together with adequate force. This should always be checked before making any alteration to the locking pressure on the hydraulic system. Before doing this the cause of flashing should always be ascertained.

Having made sure that the locking of the mould is satisfactory, attention can then be turned to the heating cylinder. It may be that one of a pair of heater bands has broken down and that the other one is over-running. This can occur where two bands are controlled by one thermocouple. A more likely and possibly more frequent cause is that a thermocouple has been pulled out of its pocket, either fully or partly, and that the heater band it controls is over-running. A still further, often unsuspected, cause of flashing is that too much mechanical heat is being put into the material because screw back is taking too long, is too fast, or is variable.

Finally, a check should be made on the injection pressure setting and the injection rate. It may be that the flow control valve has become heated and that injection is taking place too fast.

(c) *Mould at fault*. A well-designed and constructed mould will, of course, give flash-free mouldings unless the machine has been wrongly set. After a time most moulds become slightly worn and may show a tendency to flash, but the signs of this starting may be the appearance of a small burn mark—perhaps better described as an incipient burn mark—appearing at a parting line. If this develops into a bad blemish it will be necessary to repair the mould. Slight flashing at ejector pins may occur and may be acceptable but if they develop ejection becomes difficult; this, in turn, causes the cycle to be lengthened, allowing more time for plasticization in the cylinder and making the flashing worse. With hot-runner moulds, any malfunctioning of heater circuits can alter the viscosity of the melt and cause flashing.

(d) *Process at fault*. It must always be remembered that the injection moulding process involves the balancing of a number of variables in order to achieve optimum conditions. The appearance of flash is often the sign that something is out of balance. For example, a lowering of viscosity of material entering the mould causes more pressure to be exerted within the mould. This reacts against the mould locking force and if this is at its maximum, the injection pressure has to be reduced. Conversely, reducing the temperature increases the viscosity of the melt; pressure is then increased to counteract this, resulting in an increase in the rate of injection. The material is sheared more rapidly and then the pressure, acting on the more highly sheared, less viscous material, is sufficient to cause flashing. Attention to one variable only in treating flashing, as is the case with many other faults, may lead to further difficulties and it is always the best policy to try to discover exactly what is the cause of the trouble. In the end, it might simply be a small piece of a previous moulding preventing proper mould locking!

Flow marks

Material entering a mould always follows an almost identical path with successive mouldings. From a sprue gate into a flat plate it forms a spreading disc pattern. From a side gate of proper design into a flat plate it will flow in an ever-increasing semi-radial pattern. It will always flow more rapidly along wider and deeper channels and when these are full it will spread into the narrower and shallower sections. On reaching an obstruction, turbulence will result and the general flow path will be divided. The mark of the turbulence will be carried along, sometimes to the end of the flow-path. Some pigmented materials show this very clearly. Flow marks are faults which should be thought about at the very outset of mould design because the location of the gate, its type, and the position of any obstructions in the mould, will influence the type and intensity of any flow marks that form. If flow marks are inevitable in a moulding, the mould designer should try to make them appear on a surface where they will not be seen or will not cause the mouldings to be rejected. Thus, the siting of a tunnel gate perhaps near the underside of a moulding rather than near the top surface (appearance surface) might produce the more acceptable mouldings. One of the most common causes of flow marks is lettering on the appearance surface. Whether this is embossed or intaglio makes little difference. The turbulence occurs and generates a flow mark.

In general, flow marks do not affect the performance, but only the appearance of mouldings and perhaps the best thing to do, realizing their inevitability, is to mask the effect with a suitably patterned decoration.

Poor surface finish

This fault can be caused either by the process or by the mould. It can rarely be traced to machine malfunction and unless really bad material has been supplied, it cannot be attributed to that cause.

Obviously, the surface finish of a moulding cannot be better than the finish of the surface against which it has been moulded. If a good surface finish has been provided in the first place, this could be damaged by misuse or accident. In the former category are such things as mechanical damage by using hard tools for the removal of a difficult moulding, or damage by rust or by being sprayed with an anti-freeze material as is sometimes used in mould chillers. These sometimes become slightly acidic after long use and etch any metallic surfaces with which they come into contact. Even if rusting is avoided, hard water evaporating on the surface of a mould will leave a deposit of chalky material which is often difficult to remove. Accidental damage may, of course occur if the mould closes on a piece of plastic material. It may also happen if a tool being used on one half of a mould, slips and comes into contact with the other part. A protective plate will help in avoiding such an accident.

Many materials, most markedly, perhaps, low-density polythene, give the best surface finish when moulded at fairly high temperatures. The reason is obvious. The lower the melt viscosity of the material, the closer will be its contact with the mould. However, the higher the temperature, the longer will be the cooling time and the slower the output rate. Obtaining the maximum output rate may entail some sacrifice of surface finish, but so long as this is acceptable, and gives a saleable product and one which will not break down in service, everyone is happy.

Filled materials, such as nylon, acetal, polycarbonate, and polypropylene, where glass fibre, asbestos, or talc are used as fillers, always give the best surface finish when the mould is filled rapidly through a restricted gate. It seems that this treatment causes each particle of filler to become coated with a thin film of the plastic, which prevents any roughness that might be expected if the filler came to the surface of the moulding. As a general rule, the shear rate/viscosity curve for mineral-filled plastics shows the lowest viscosity at the highest shear rates and is much steeper than for unfilled materials. Obviously, the lowest melt viscosity gives the best finish.

Scuffing

Sometimes when a mould opens, the vertical wall of the moulding rubs against the wall of the cavity and receives a scuff mark in the direction of opening. The most obvious cause of this is that the mating surface of the cavity has become 'hobbed', possibly by too rapid closing of the mould, and the edge is burred. If the mould has not been damaged and scuffing occurs, it may be caused by excessive packing of the mould during injection or by the draught angle on the mould walls being too small. With excessive mould packing, the cure is to adopt the correct mould filling procedure and to reduce the hold-on pressure and time (or possibly to dispense with hold-on altogether). Scuffing caused by over-packing will nearly always be associated with residual stresses in the moulding, which can cause warping or breakdown in service.

The draught angle for moulds to enable articles to be produced free from scuffing depends on the shape of the component and on the material. As little as 0·5° has been used satisfactorily for polypropylene, and sometimes no draught angle at all can be permitted. For materials such as ABS and acrylics, a more generous draught is required. Even if the draught angle is adequate, there may be some difficulty in extracting mouldings without causing scuffing if the mould has been polished at right-angles to the direction of mould opening.

Short mouldings

The causes of short mouldings are many and varied and are sometimes difficult to diagnose. It is, for example, perfectly easy to flash a mould and still obtain a short moulding. On the basis that the same property, too low a melt viscosity for the material, will probably be a cause of flashing and that it also ensures easy mould filling, the combination of both faults should be impossible.

The failure of a mould to fill is brought about by a combination of factors, the most important of which is the build-up of back pressure in the mould before it has completely filled. Commonest among causes of this build-up of back pressure is air or gas trapped in the mould. In the event of complete trapping of air, that is, it cannot escape at all, a burn mark is inevitable if some mould filling has taken place. However, air trapped in the mould but being slowly vented may not cause oxidation but may prevent the material from reaching the limit of its travel in the mould before it solidifies. Thus, short mouldings may develop into burn-marked mouldings if the vent becomes completely blocked. This can be the result of a new mould bedding down—new moulds need just as careful 'running in' as new cars—or by grease or dirt entering the vent. In this context, it should always be remembered that the steel of which the mould is constructed is elastic, and under the high pressure during locking it can be compressed quite considerably; in fact, enough to close small venting gaps.

Back pressure is also built up in the mould by solidification of material as it enters and flows through the mould channels. This has been discussed in Chapter 6 and it should be reiterated that if the gap between the walls of the mould is going to be reduced by more than 50 per cent during filling there will be a strong possibility of producing a short moulding. In these circumstances the operator probably increases the injection pressure. A full moulding may then be made, but it will contain residual stress near to the gate. On reflection, it becomes obvious that what is needed is more rapid filling, requiring, as previously indicated, a hydraulic system capable of delivering the required amount of hydraulic fluid in the time allowed for filling.

By now, it will be appreciated that some criticism of the standard fault-finding chart is justified. To indicate the various expedients to avoid short mouldings without discussing the underlying questions can lead to a trial-and-error situation in which the operator is virtually enjoined to 'try everything and see if it works'.

Eliminating short mouldings, therefore, involves the proper energy being available in the machine hydraulics, the material being in the correct state of fluidity, and the mould filling without causing back pressure of the kind that blocks the way through it. If short mouldings are made and all the obvious things such as feed setting, temperatures, and power requirements are in order, then one must look further for the cause and its cure.

A broken, or ineffective back-flow stop valve can cause starving of the mould by reducing the available pressure and feed. Adjusting the feed to compensate will not give a complete answer because the leakage along the screw will be reducing the effective filling *rate* and short mouldings will result. Malfunctioning of heaters and temperature controllers may sometimes cause short mouldings, but this should easily be diagnosed. A blocked injection flow control valve should similarly be noticed because the injection rate will have fallen off. A blocked nozzle should also be quickly diagnosed.

It is also assumed that runner and gate systems have been properly designed and executed. In a new mould it may be necessary to increase gate dimensions, but this should always be carefully carried out. A well-tried mould should need the gate altered only for a different grade of material. Altering a mould by putting in two or more gates rarely has a beneficial effect and sometimes causes more problems than it solves. The general principle of gating in only one place applies to nearly every possible situation.

A technique that can be of great help in assessing how a mould fills is that of deliberately producing short mouldings. Having set the appropriate filling rate and checked that the gate size and all temperatures and pressures are correct, a series of short mouldings is produced, adjusting the feed control to give the varying degrees of shortness. As the trial progresses, a study of the mouldings will indicate exactly how the material is flowing and whether, for example, additional venting will be required for complete mouldings to be produced.

If the mould-filling calculations are correct and the machine conditions are set accordingly, the temperature of the mould will make little difference to mould filling and the making of short mouldings except in special circumstances. An example of such a case concerns the moulding of a sheath on a solenoid, the material being glass-filled nylon. The mould in this case needed to be at between 80 and 90°C to mould a complete sheathing.

Sink marks and voids

Sink marks and voids are two manifestations of the same fault and their complete eradication is extremely difficult. Diagnosis is easy. All plastics take up more volume when molten than when solid, and where there is more volume there will be more shrinkage; in other words, thick sections shrink more than thin ones. In any part of a moulding where there is a sudden change in section which locally increases the volume of material, either the surface will be drawn in as the material shrinks, giving a sink mark or, if the surface has become hard enough not to sink, the shrinkage will take place within the body of material causing a void. Overcoming this problem is not simple.

One method of reducing or preventing sink marks and voids often recommended in manufacturers' literature is to increase the injection holding time (plunger dwell). This may have some effect, particularly where thick mouldings are being made using a large gate, but there is a danger that the pressure will not be evenly distributed throughout the material in the mould and that some parts will be more highly stressed than others. Thus, a moulding may seem to be perfectly free from blemish, but would break down in service under adverse environmental conditions. This technique demands considerable care in making sure that the holding pressure is not so great as to cause undue stresses.

Another recommended method is to reduce the temperature of the material in the cylinder (on the basis that cooler material shrinks less). This can have only a limited effect because it is extremely unlikely that a reduction of more than a few degrees could be tolerated, bearing in mind that, as discussed in an earlier chapter, moulding should always be carried out at as low a temperature as possible. The average volume change on heating for crystalline plastics is about 1 per cent for every 10°C of temperature change (about half that figure for amorphous materials). If a reduction of 10° could be tolerated this would make a difference in shrinkage of only 1 per cent when the effect being considered is the result of a 14–18 per cent change in volume. Furthermore, the lowering of temperature would also cause a reduction in the ease of filling because the viscosity of the melt would be somewhat higher. Consequently, more pressure would be needed, bringing with it the possibility of greater residual stresses.

Rapid mould filling with the correct amount of feed at the lowest possible temperature is the most effective way of reducing sink marks and voids in the great majority of cases. The main exceptions are very thick sections injected through very large gates, using low-viscosity materials (with the proviso above). Although these faults are difficult to eradicate completely, it is sometimes possible to make a virtue of necessity and to design into a component a feature that will actually make use of the fact that a sink mark may occur. Thus, if a thicker section is desirable it may be possible to make an indentation in the moulding at, or near, the place where a sink mark may occur.

If the design of a component renders it virtually impossible to mould without either sink marks or voids, the former should generally be chosen. Voids are always likely to provide centres where notches can occur, thus reducing the impact strength of the moulding. By adjusting conditions, voids can often be changed into sinks marks, and so long as the sink mark is not detrimental to appearance, it may be less harmful to performance than a void. It may be possible to mask sink marks by designing in styling beads or to change the contour where the sink mark is likely to occur. In this event, the styling should not be of such a nature that it turns a sink mark into a void.

Sink marks and voids are perhaps the most difficult faults to deal with satisfactorily. They are basically a problem of design, but if they have not been considered at the design stage, the only advice that can be given to the moulder is, ‘do the best you can’.

Splash marks, mica marks, silvering

This fairly common fault is easily recognized by the streaky, silver appearance of the moulding, nearly always radiating from the gate area. The cause is the vaporization of volatile matter in the advancing front of material as it enters the mould, or sometimes within the mould itself. Volatiles can be the products of decomposition or depolymerization of the material or may be due to the absorption of water by the plastic prior to moulding. The volatile material entering the mould condenses on the surface, forming liquid droplets. As the material enters the mould, the liquid droplets are carried along, forming the streaky pattern. The same effect is obtained if the mould is cooled to below the

Fig. 8.2 Splash marks (silvering or mica marks)

dew-point of the moulding shop and atmospheric moisture then condenses on it, producing the water droplets. This is illustrated in Fig. 8.2.

The cure for splash marks is, therefore to eliminate all volatile matter. If the material is damp, it should be dried; if the volatiles are being caused by decomposition or depolymerization, the cylinder temperature should be lowered, and if moisture in the mould is to blame, its temperature should be raised.

Materials most likely to be affected by moisture are those that contain hydrophilic groups in their molecular structure, especially a $\rangle C{=}O$ group (carbonyl group). Among the commonly used plastics, acrylics, nylons, and cellulosics are the most hydrophylic and therefore the most prone to splash marking. However, owing to the presence of finely divided carbon (carbon black), some out-door weathering grades of polyolefines have a tendency to absorb moisture, and others, such as ABS and polycarbonates, tend to attract sufficient moisture to cause splash marking by adsorption of moisture on to the surface pores of the granules themselves.

Manufacturers normally supply material in the dry state. Only rarely is the moisture content too high to permit satisfactory moulding, but whereas a small quantity of moisture (less than 0·1 per cent) can often be tolerated in nylon or acrylic material, *no moisture at all* can be tolerated in ABS and polycarbonate if splash-mark-free mouldings are to be made. These must be dried immediately before use and put hot into the feed hopper. Nylon and acrylic should be stored in air-tight containers or polythene film sacks and re-work should be kept dry. Details about drying equipment and conditions are discussed in Chapter 11.

Decomposition as a cause of splash marking is not easily recognized unless the purgings from the machine nozzle are examined. If there is any tendency towards bubbling or frothing, the cylinder temperature should be lowered and a check made to ensure that there are no hot spots. Excessive mechanical heating developed during screw back can also be a cause of overheating and should be suspected as the cause of splash marking if any variations in screw-back time have been observed prior to the fault developing.

Some moulders believe that splash marking is caused by shear heating in the gate, basing their assumption on the fact that splash marks nearly always radiate out from the gate. With a well-designed gate system this is extremely unlikely for the reasons already discussed in the section of this chapter dealing with burn marks.

Splay marking

This fault, which appears on a great many mouldings, shows an area, generally near the gate, which has an uneven, lined appearance. It may be reminiscent of the grooves on a gramophone record or it may appear simply as an irregular pattern of lines, roughly in a half-circle round the gate. The reason for splay marking is slow pressurization of the last part of the mould. As the back pressure in the mould builds up, the rate at which material flows through the gate is reduced and the molten plastic does not have a low enough viscosity to give good conformity to the mould surface.

To overcome the defect it is important to fill the mould rapidly at one attempt. If the machine runs out of power before filling is complete the last material can enter the mould only at a very much slower rate. Pressure will gradually build up in the hydraulic system and will be transmitted through to the material entering the mould, forcing a little more in. The material already there will have cooled, and conformity to the mould shape cannot be obtained. When splay marking occurs, therefore, the first action should be to check that the hydraulic system will enable filling to take place at the correct rate and in the time that can be allowed. If it is, and splay marking still persists, the sprue and runner system should be examined for any points where pressure and flow could be lost during filling and, if necessary, made larger or more direct. The land length of the gate should also be made as short as possible.

Splay marking is particularly liable to occur in mouldings made in acetal and this may be due to its somewhat higher melt viscosity. As already mentioned, however, the fault has been observed when small mouldings in many other materials are produced on hand machines or on low-power machines. There is little that can be done to improve such mouldings, and it must be remembered that their performance may be impaired. Any moulding showing a discontinuity of surface finish is suspect and in aggressive environments may be expected to crack or break down completely.

Staining

Occasionally, a highly polished mould surface suddenly takes on a dull, discoloured appearance and this shows up on the moulding. There can be several reasons for this: decomposition of an additive, or the separation or blooming of a pigment (that is, the migration of the pigment to the surface of the plastic), or simply slight oxidation of the material, not sufficient to cause a burn mark but liberating a little gas which reacts with the mould surface.

The cure for this fault depends on the material being used. If there is a percentage of re-work, there may be contamination, for example, with mould release agent. There might also be a quantity of slightly decomposed or depolymerized plastic. Dry-colouring pigments may have become damp, or there may be preponderance of fine, dusty material. With PVC, decomposition causes the liberation of some free hydrochloric acid that can quickly stain a polished mould. All these matters need to be considered and eliminated in turn until the cause is discovered. It is nearly always a material fault, but sometimes, by altering conditions slightly so as to reduce the possibility of decomposition, a cure can be effected.

Sticking

Mouldings sticking in the mould can be a fault of material, machine, mould, or process. In simple terms, it is caused because the moulding is too big for the mould and will not shrink enough to allow ejection, or because it shrinks too much on to a core or other part of the mould having a minimal draught angle.

If the material is to blame, it is usually a matter of the wrong grade being used. Too free a flow will cause over-pressurizing of the mould and is rectified by reducing pressures and temperatures. If the material is too stiff, the operator will increase the injection pressure and then sticking will probably take place in the sprue bush, or any other thick sections that have become over-pressurized.

Any failure of the machine that allows undue pressure to be applied can cause sticking, and one of the most common situations is that which has already been described for splay marking. If the power of the pump is not great enough, mould filling will be difficult and will take longer than desirable. To ensure a moulding free from shorts, sinks and voids, a high holding pressure will be applied. In this respect, the difference between pressure and rate of injection is significant. It may be possible to build up to a very high pressure, but if it takes too long a time and the rate of injection is low, sticking in some parts of the mould is inevitable. Similarly, too slow a rate of filling causes more severe orientation with its attendant increase in the amount of shrinking, and this may make it more difficult to extract a moulding from a core.

A mould which has been badly designed will often give sticking problems. Rough, unpolished parts, tapers not acute enough, and even slight blurring of the edges of the mould, can cause the moulding to stick in the mould and make ejection difficult.

There is always the possibility that process conditions may contribute to sticking. Slowing the injection rate to keep the gate open longer and so reduce sink marks on thick-section mouldings, can often lead to a final build-up of pressure at the end of injection, particularly near the sprue and so cause sticking. Setting the feed too great and maintaining a long hold-on period can have the same affect.

As with splay marks, sticking mouldings should be slightly suspect. They may appear to be satisfactory (although stress whitening may develop during ejection), but having been over-pressurized there is almost sure to be some residual stress which will lead to cracking or crazing at a later stage.

Warping

Mouldings warping after removal from the mould are very common and are always the result of frozen-in stress, differential shrinkage or, with filled materials, filler orientation. This is nearly always a processing problem, but it can also be caused by a machine failure or incorrect component design. Anything that prevents injection at the correct rate will contribute to warping. Thus, the non-return valve on the screw not seating and allowing back flow along the screw flights prevents proper injection and causes warping. In fact, a moulding which would have warped, except that it was restrained by a strong rim, such as the base of a bucket or bowl, will have a tendency to break down in service. Expedients such as differential mould cooling and jig cooling are only partial palliatives. The only real cure (given correct component design) is correct injection and mould filling.

Both the two expedients mentioned will certainly produce unwarped mouldings but the question is: how long will they retain their shape? For example, a component for use at elevated temperatures may be free from warping when first made, but after a few weeks or months of use at its normal working temperature it may warp sufficiently to become unusable.

When warping occurs, the injection rate should first be checked. Short mouldings should then be made by the method described in the section on Short Mouldings, to determine how the mould is filling; and then, having ascertained that filling is uniform and that material is reaching all extremities of the moulds at the same instant, mouldings should be made at the same (correct) injection rate with exactly the correct amount of feed. In nearly all instances, this method will prove satisfactory. If the material does not flow uniformly in the mould it will be necessary to ease some of the channels or, with a deep-drawn symmetrical article, to true up the mould parts to make sure that all the flow paths are of equal thickness.

If warping is still a problem, even with this moulding technique, the mould may need to be altered so that several pin-point gates can be used to cause virtually undirectional flow during filling. On account of the cost of altering a mould, this should be thought of only as a last resort if all else fails.

Fig. 8.3 Typical weld line

Weld lines

Weld lines occur in mouldings where two or more streams of material meet and fuse together (Fig. 8.3). At best, the weld may be a small, barely discernible mark; at worst, the component breaks at the weld line. They are difficult to eradicate and are not caused by any deficiencies in material, machine, mould, or process, but are inherent in the design of the component. Modifications to the mould and the use of a freer flowing grade of material are ways in which the effect of weld lines can be reduced, but the chief factor involved is finding the optimum processing conditions. 'Jetting' (Fig. 8.4) is a special type of weld line caused because the jet of material entering the mould freezes before it welds together.

Fig. 8.4 Typical jetting

With regard to the mould, it is seldom advantageous to feed from two or more gates. This may assist concentricity but will almost certainly produce a weld line at right-angles to the line joining the two gates. Side gating, where there is a single hole in a component, will nearly always produce a weld line at the side opposite to the gate. When a moulding has several holes, a diaphragm gate in the largest will help to reduce the effect, as will over-flow wells beyond the weld line, but each smaller hole will produce its own pattern of weld lines. Mould temperature control helps to reduce the effect and using the maximum feasible temperature will often almost eliminate them. However, with crystalline plastics, a high mould temperature not only increases the crystallinity but also increases the size of the crystals, and it is possible to end up with a brittle component having even less strength than one with weld lines.

The most successful way of tackling the problem of weld lines consists in trying to arrange so that the two streams of material meet while their leading fronts are still capable of fusing together. It involves a suitable combination of mould and cylinder temperatures, rapid injection and mould filling, and making sure that there is adequate mould venting. Obviously, anything that hinders the flow of material in the mould increases the severity of weld lines, and sometimes even the judicious application of a little mould release agent will help by lubricating the mould surfaces. It also helps if polishing marks on the mould surfaces are in the direction of the flow of material. As with warping and short mouldings, the deliberate making of short mouldings will help in showing why the weld lines are developing. Occasionally, weld lines can be made to form in positions where they are unobtrusive or where fracture will cause the minimum of inconvenience. This involves careful control of the

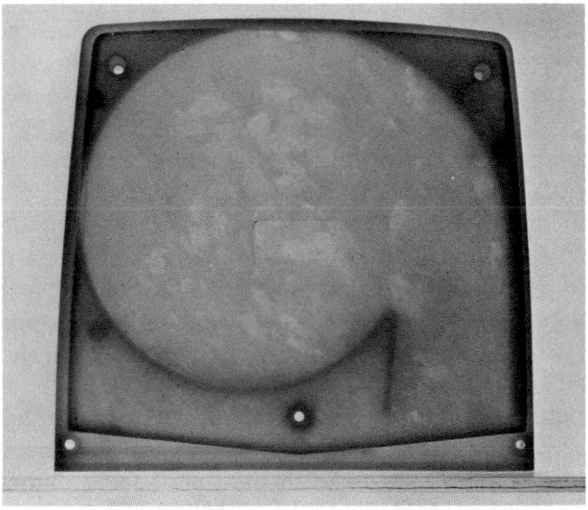

Fig. 8.5 Windows in a moulding

injection rate, bearing in mind that the injection flow control valve may become warm during the moulding programme and alter its characteristics.

Windows

The formation of windows—areas of amorphous material in an otherwise largely crystalline mass—has already been discussed, and all that needs to be said further is that they will not appear where plasticization is thorough and complete. They may be produced in small machines being used for the making of mouldings of about the maximum capacity of the plasticizing cylinder, in which case a longer time cycle will generally effect a cure. Windows should be avoided because they provide areas of discontinuity and have the same effect as notches in reducing impact strength (Fig. 8.5).

Witness marks

Witness marks are always associated with badly mating mould parts. This may be caused by the mould not locking properly and allowing flash to form and accentuating the mark. They are sometimes masked by the inclusion of a styling bead and, of course, are less apparent on patterned mouldings than on highly polished ones.

Other faults and problems

The above collection of moulding faults, although fairly comprehensive and covering most of the day-to-day problems that occur in the moulding shop, may not cover some of the more esoteric difficulties encountered. Nevertheless, the same approach to solving such problems, that is, the analytical approach, should be of assistance.

One of the most prevalent problems encountered in injection moulding is the variation of the process. Many workers have observed variations in moulding weight from shot to shot of 2 per cent with a maximum spread of as much as 6 per cent. Obviously, with this amount of variation there must be something badly wrong with the machine and it is a sad commentary on the state of injection machinery in the United Kingdom that it appears to be the norm. In well-maintained machines, a variation from shot to shot of no more than 0·2 per cent can be expected, any greater changes being caused by actual changes in moulding conditions. The causes of variations in shot weight and their relationship to the machine and mould operation are shown in Table 8.1. Attention to the matters mentioned in the right-hand column of the table should assist in obtaining better reproducibility and less variation in the process.

Table 8.1 Injection moulding process variables and their causes

Variable	Possible cause
Injection pressure	1. Dirty oil; blocked filter 2. Faulty pump; cavitation 3. Faulty relief valve 4. Overheated or cold oil
Injection rate	1. Overheated oil 2. Air, water, or sludge in oil 3. Faulty control valve 4. No accumulator, and pump too small for required rate of flow
Melt temperature	1. Heater band malfunction 2. Thermocouple malfunction 3. Faulty temperature controller 4. Hot spot in cylinder
Shot volume	1. Faulty switch on feed control 2. Worn cam follower on feed control 3. Worn and leaking check ring (back-flow stop valve) 4. Incorrect screw-back speed 5. Malfunction of pressure reduction switching 6. Nozzle dribble 7. Leakage at nozzle/sprue bush
Mould temperature	1. No proper mould temperature control 2. Dirt or rust in cooling channels 3. Cooling channels wrongly connected
Mould characteristics	1. Mould incorrectly set 2. Mould distorted 3. Mould badly vented 4. Mould badly constructed 5. Mould badly designed

To sum up the discussion of the causes and cures of faults in injection moulded components, Table 8.2 should be consulted. Here, the main faults are listed on the left. In the second column is a generalized cause, and possible remedies are listed on the right, together with the appropriate text references. Such a table should be used with discretion; nothing is more valuable in the diagnosis and cure of faults than a thorough knowledge of the process and the ability to apply that knowledge to the problem in hand. There is no short cut by means of tables, charts or algorithms because, sooner or later, a fault will be encountered that no one has ever mentioned before.

Table 8.2 Faults in injection moulding

Fault	Generalized cause	Possible cures	See page
Short moulding Sinks and voids Weld and flow marks Poor surface finish	Cavity not filling properly	Increase injection pressure	136
		Increase injection rate	138
		Increase cylinder and mould temperature	134, 141, 144 135
		Increase feed	120
		Improve mould venting	95
		Increase size of gate, sprue, and runner	96 ff
Burn marks	Lack of mould venting	Improve mould venting	130
Mechanical sticking Flash Warping	Over-pressurization of cavity	Reduce injection pressure	136 & 142
		Use minimum hold-on pressure and time	134
		Fill more quickly	143
Silvering and splash marks Decomposition	Overheating of material	Reduce cylinder temperature	139
		Reduce effects of mechanical heating	131
Distortion	Ejecting too hot	Increase cooling time	133
		Eject over larger area	133
Jetting	Linear velocity of melt too high	Reduce velocity or break up stream of melt	144
Lamination, crazing Surface cracking	Frozen-in stresses	Ensure rapid filling at lowest possible pressure	132

9

Housekeeping as a factor in efficiency

It does not always follow that the cleanest factory is the most efficient, nor that the dirtiest one is the least. It is, however, quite clear that good housekeeping aids efficiency and that bad housekeeping can be detrimental to profitability, to safety, and to health. At the present time, when job satisfaction plays such an important role in industrial management and relationships it would seem that clean and pleasant working conditions have their part to play. A person occupied in keeping the moulding shop clean may be just as gainfully employed as a machine operator, but there is also a great deal to be said for encouraging all persons working in the shop to have good housekeeping in mind. With an individual fitter/technician or a team of fitters made responsible for a group of machines, competitions can be held for the best kept section; and this, with an attendant productivity bonus, certainly has the effect of increasing overall efficiency.

Cleanliness

Not only is a clean machine in clean surroundings easier and more satisfactory to work on; it enables the operators, setters, and maintenance staff to see and rectify troubles before they reach unmanageable proportions. Ideally, working a 120 h weekly shift scheme, the time from 6 a.m. or 8 a.m. on Saturday until midday should be devoted to cleaning and minor maintenance jobs. If this cannot be arranged, cleaning and maintenance staff should attend to machines as they become idle at the end of a project, before setting up a new job.

(a) *Dirt, dust, and granules*. Some of the treatment that injection moulding machines receive is not calculated either to improve performance or to prolong the life of the machine. Starting where the material starts, in the feed hopper, there is often a spillage of granules or powder on to the machine when filling the hopper. This falls on to the hot heater bands around the cylinder and tends to stick. After some time, virtually the whole of the cylinder begins to look in poor condition. Powder and dust fall on other parts of the machine, settling on cams and microswitches and penetrating into cracks and spaces. If not removed, this dust can penetrate into other working parts of the machine and, where the oil tank is not totally enclosed, it has been known for plastic granules (and even mouldings) to be found in the oil.

(b) *Purgings*. Purgings from the cylinder are another source of trouble. If left until solid on the shelf below the nozzle they may be difficult to remove, and where an unattended machine is allowed to dribble at the nozzle by being left with heaters switched on, there is the possibility of wires becoming encased. While hot, this can be removed, although with difficulty, but when allowed to cool a hacksaw is needed to free the wires, and not without some hazard to the safety of the insulation. Sometimes the weight of plastic is great enough to pull thermocouples out of their pockets and to cause wires to break when the carriage is moved.

(c) *Disposal of mouldings etc*. Especially when starting up a machine or after fitting a new mould, a considerable number of mouldings have to be made and examined. Often there is nowhere to put them except on any flat surface of the machine, which quickly becomes festooned with mouldings in various stages of completeness. If a table or trolley is provided, the danger of leaving the mouldings on the machine is avoided and a step is made towards a tidier shop. Tools, nuts and bolts, mouldings, purgings, and even cigarette ends all get mixed up on top of the machine, and some even end up in the granulating machine.

This sort of situation can rapidly develop unless a constant watch is made on housekeeping and cleanliness. Who, for example, has not seen odd nuts and bolts, washers, and perhaps even a spare nozzle put inside the control cabinet of a machine 'for safety'. The provision of proper tool cabinets on trolleys and a spare parts box attached to each machine avoids this sort of trouble.

(d) *Care of packages*. Most plastics, except those provided in bulk, are packed in paper sacks, generally with a polythene film interlayer. Although fairly tough, the occasional sack ruptures and a trail of granules on the floor marks its travel through the moulding shop. This should be avoided, if only to minimize waste, because of the cost of the material. The dust and dirt and the hazard of granules on the floor can all be avoided by using adhesive tape as soon as the tear in the sack is seen.

(e) *Granulators*. A frequent cause of dirt in the moulding shop is the small granulator placed by the machine so that the operator can de-gate the mouldings and re-grind scrap on the spot, ready for recycling. There are two objections to this, and to the practice of drilling, machining, or otherwise treating mouldings while waiting for the next cycle. These are: (1) dust is generated near to the machine and this, as has already been explained, can be detrimental to the control system; (2) by recycling scrap in this way a true material balance and machine utilization is difficult to obtain.

(f) *Lubrication*. Turning to the machine itself: containing oil, circulating oil, and being lubricated with oil, the machine and all its surroundings all too often become an oily mess. Oil on the floor is a hazard, making it slippery and attracting dirt and dust; oil on the tie rods or too much oil and grease around the machine makes mould setting and maintenance very dirty work. If mechanically actuated lubrication systems are set to give too much lubrication, all parts of the machine where oil can accumulate become awash and it is then very difficult to keep mouldings clean. Although lubrication is important, so is the cleanliness of the machine.

(g) *Coolant fluid and hydraulic oil.* Another hazard to cleanliness in the factory is the cooling system. When moulds are disconnected from the temperature controller there is nearly always some spillage of water or coolant. Operators should be trained to use a bucket or container to catch the water and to drain the mould, using a low-pressure air line to clear all the channels in the mould. Water drained from a mould often contains some rust and if an antifreeze material is included in the coolant, it should be saved and returned to the system after filtering.

Leaks from the hydraulic system, particularly from flexible hoses, are another possible source of dirt in the moulding factory. They should not be allowed to develop because, if a flexible pressure line fractures, the whole moulding shop can be showered with oil.

(h) *General cleaning.* Often overlooked, but nevertheless a fruitful source of dust and dirt in the moulding shop and a common cause of contamination of material, is the dirt lodging on roof supports, beams, pipelines and electrical trunking. On a dry, windy day in summer with windows and doors open and ventilation systems working, the factory can soon be full of air-borne dust. Regular cleaning can reduce this hazard and small birds can be excluded by means of wire screens. If birds build their nests in the factory roof they become a very mobile source of dirt and contamination.

Contamination and waste

(a) *Keeping material clean.* The importance of cleanliness in the moulding shop is nowhere more apparent than in the making of mouldings free from contamination. Safeguarding the material in the sack is an obvious matter, but is often ignored. Partially used sacks or containers should be re-sealed or covered and raised off the floor, in case of water or oil leaks. Material stored by the machine for use during the following shift should be raised on a pallet and covered, in case dust or dirt collect on the outside of the sacks. It is very easy to tip dirt or foreign granules into the feed hopper from a dirty sack. The lid should also be kept on the feed hopper. It often is not!

It is not always realized how difficult it is to get rid of black specks in white mouldings (or white specks in black ones). One single black granule in a hopper full of white material can contaminate as many as half the mouldings that can be produced from the full hopper. With strict inspection, this can mean a great loss of production time and material.

(b) *Protecting machines and granulators.* For critical jobs almost clinical cleanliness is desirable and it is *absolutely impossible* to make good mouldings in a light colour if a granulator working on black material is operating in the vicinity. The high static charge which all dry plastics carry attracts dust, and even the mould, if warm and dry, will attract statically charged dust. The only safe course when working to critical limits is to isolate the machine by means of plastic film screens on a light framework and to blow in clean air to maintain a slight positive pressure inside the screen. The alternative is to do away with all granulators and other sources of dust within the moulding area, or to enclose the granulators in dust-proof (and noise-proof) boxes.

(c) *Care of re-work*. Scrap mouldings, sprues, and runners that are to be re-ground and re-worked must be kept free from contamination. They, too, take up a static charge and attract dust and dirt. Some plastics also attract moisture and in a few days, even with a relative humidity as low as 50 per cent, will become too damp to mould. The same care should be exercised with re-work material as with the virgin plastic in keeping it free from contamination.

(d) *Air lines as a hazard.* A very common cause of contamination in the moulding shop is the compressed air system. It may, of course, contain moisture which is itself a hazard. It is more than likely, however, that the main use of the compressed air line will be for cleaning out feed hoppers and the feed throat of the machine. Of course, the plastic or dust and dirt must go somewhere and once air borne, it may land anywhere. The most damaging place is in the feed hopper of an adjacent machine. This use of compressed air is not to be recommended. It is much better to use a vacuum cleaner, either one of the powerful industrial models or even the small hand machines used in the home and produced by several of the domestic appliance manufacturers.

Moisture in compressed air can also contaminate mouldings if it is used for air ejection. The moisture can settle in the mould to produce splash marking, but there are often traces of oil in the air which can make mouldings quite dirty. It is wise to make sure that the compressed air is clean and dry when used in this way by the use of a sump and filter.

(e) *Getting rid of contamination*. A great deal of needless waste often occurs in getting rid of contamination. It is pointless to continue moulding if the components produced will not pass inspection. The contamination may be in the feed hooper or in the bag of material used to charge it, and the first action is to discard this material. The procedure is similar to that recommended for colour changing: first empty the feed hopper (retaining the material for later examination); continue to mould until short mouldings result; screw back as far as the screw will go (this depends on the amount of material left in the screw flights); purge the cylinder to air; thoroughly clean the feed hopper and the feed throat; put a small quantity (a few handfuls) of material in the feed throat; screw back and purge; repeat once or twice or until satisfied that no more dirt is coming from the cylinder; re-fit the feed hopper, put in a *small* quantity of material and start moulding again. If the mouldings are satisfactory, the feed hopper can be filled and the lid put in place; if not, the process is repeated.

This method of getting rid of contamination from the cylinder uses the minimum of material and avoids waste. Failure to clean the cylinder in the manner described, for example, by just emptying and re-filling the hopper with clean material and then continuing to mould, can result in the making of dozens or even scores of contaminated mouldings.

(f) *Optimizing production efficiency*. Dirt and contamination of material is not the only cause of waste in the moulding shop. It can also result because not enough thought has been put into the best method of doing a job. It may be argued that this is nothing to do with housekeeping, but if that concept is considered in its broadest sense as the elimination of any factor in operation which is detrimental to profitability, a case could be made for discussing it here.

The kind of project concerned involves the moulding of very small articles, particularly in heat-sensitive materials, such as acetal and nylon. By normal methods, using a sprue and runner system, the situation often occurs where, for every 20 g of mouldings there may be 40 g or more of sprue and runner. This is not wasted, but the cost of re-granulating and mixing with virgin material must be taken into account. If the material has become at all oxidized during its passage through the machine, it may not be possible to use it for the job in hand and it then has to be sold as scrap for between 10 and 30 per cent of its original price.

Very often, size of machine available determines how the job is to be done, and this was the case in the production of the tyre valve caps referred to in Chapter 5. They were required at the rate of 30 million per annum. A 40-impression mould was made in which the runners weighed 40 g and the mouldings 20 g. The cycle time was 30 s (determined by the need for the runners to be hard enough to eject) but if the caps could have been moulded without runners the cycle time would have been between 1 and 2 s. Two single-impression moulds could easily have produced the annual requirement, but in that period 45 tonnes of material were moulded and 30 tonnes had to be re-ground using the 40-impression mould. If they had been moulded without re-work, only 15 tonnes of material would have been processed. In deciding how best to do a job, the side issues involved may be more decisive than the actual job itself. In the above case, the cost of the material used was not just the buying price per tonne but that cost plus the cost of re-grinding 30 tons of material at, say, £50/tonne.

Safety aspects of housekeeping

It would be fairly safe to say that, if all aspects of good housekeeping were strictly observed, there would be very few accidents in the moulding shop. Unfortunately, good housekeeping is often sacrificed in the cause of keeping production going, sometimes with disastrous results. Spillage of granules on the floor has already been mentioned, along with leakages of oil. The combination of granules, oil, and water—often met in the moulding shop—on the floor can be almost lethal. Oil spillages are sometimes soaked up with rags and sawdust which are then left on the floor for hours, or even days at a time. It only requires a cigarette end or a spark to start a serious fire. Proprietary materials are available for soaking up oil and rendering it almost non-inflammable but it really is best not to spill it in the first place. Early attention to leaks and the use of a drip tray during maintenance work will do much to avoid this hazard. If an accident does occur, resulting in a spillage of oil, it is best to soak it up, clear the floor, and wash with a suitable detergent solution. Observation indicates that very few floors in injection moulding shops are ever washed!

(a) *Flexible pipes and leads.* One often sees flexible air, water, and electrical connections draped in festoons around the machines. This cannot be thought of as good housekeeping and could easily be a safety hazard. A forklift truck or a mould lifting gantry can easily foul the pipes and wires. At best, there might be a water leak with not much danger, but a fractured electrical lead could cause a severe or even fatal accident. All flexible leads should be clipped into neat positions and mould temperature controllers should be arranged as near to the machine as possible. Obviously, if any of these connections can be run over on the floor by trucks and barrows there is an additional hazard. Even treading on such leads could be the cause of trouble. It is a good plan to colour-code all auxiliary connections to the machine with distinctive markings so that they can be easily and quickly traced.

(b) *Machine guards.* Machine guards are particularly important. If there is any possibility that hot plastic could be sprayed over an operator there should be a guard made of a robust transparent plastic such as Perspex acrylic sheet or polycarbonate sheet. These materials can be shaped quite readily in most small workshops. Expanded metal guards which are often fitted to machines are no real safeguard if hot material is being sprayed out. Hot water, hot oil, or glycol circulating through a mould can present a safety hazard and frequent checks should be made to ensure that flexible connections to the mould are not being abraided during mould opening and closing. The sheet plastic guard already mentioned will help to reduce the danger should one of the leads be fractured.

(c) *Disposing of purgings and hot material.* Good housekeeping dictates that purgings and hot material should not be allowed to stay on the machine. They can cause unpleasant burns which take a long time to heal. Remember that plastics have very high heat capacities and keep hot for a long time. Purgings of acetal are best put straight into water, especially if they are decomposing. Although unpleasant and toxic in large doses, formaldehyde, the decomposition product of acetal plastics is lachrymatory and is liable to drive the personnel out of the factory long before it can do any lasting damage.

(d) *Ventilation.* Ventilation is an important aspect of safety in housekeeping. Some plastics contain additives that can have a slight toxic effect and some might contain minute quantities of suspected materials. One such plastic is PVC. Stringent precautions are observed in its manufacture because of the suspicion that the monomer VCM might have certain physiological effects. There is virtually no chance that VCM could be liberated during the injection moulding of PVC but no risks should be taken and a special extractor system is very desirable over machines used for this purpose. All moulding shops should be very adequately ventilated, not simply by the opening of windows and doors but by means of extractor fans in the roof above the machines.

(e) *Lighting*. Adequate and efficient lighting is of importance, particularly from the point of view of safety. Lamps, tubes, and reflectors need regular cleaning and faulty lamps should be replaced as soon as possible, even though they may be in inaccessible places. Of course, faulty switches should be taken out of service immediately and replaced as quickly as possible. Although as much energy as possible should be saved by switching off unnecessary lights, this should never be done to the extent of jeopardizing safety. The desirable levels of lighting for various industries are laid down in factory legislation and should be strictly adhered to.

(f) *Safety glasses*. Some operations which take place in connection with the moulding of plastics require the provision of safety glasses to be worn when using grinding equipment, etc. It is also desirable to provide them to operators working on granulating machines because the small particles which sometimes tend to fly out of the feed throat can be harmful to the eyes.

(g) *Lifting tackle*. Lifting tackle, including slings, A-frames, hoists, and fork-lift trucks, needs to be regularly inspected to ensure continued safety in operation. Each item should be clearly marked with the safe working load, and all moulds, especially larger ones, should have their weight stamped on them. Accidents with fork-lift trucks are possible if too heavy a load is slung from a position to far along the fork. Each mould should, for preference, be supplied with its own lifting eye and make-shift slings of any description should never be used. It is unwise to sling any mould with a piece of ordinary rope and the proper equipment should always be used.

As an additional safeguard where heavy pieces of equipment, such as moulds, are being regularly handled, it is wisest for the operator to wear safety shoes or boots. These have a strong steel toe-cap and can be instrumental in saving broken bones in the feet if an accident occurs.

(h) *Ladders, scaffolds, etc*. Ladders and scaffolds need to be of robust and firm construction. Some of the portable steps often seen in moulding shops are somewhat unstable, having castors and spring-loaded steadies which come into contact with the floor when anyone mounts the steps. Until the weight is actually on the steps they are free to move, and this can be disconcerting to the operator carrying a 25 kg sack of plastic granules on his back. Strong wooden steps, suitably designed for each machine, having a hand-rail, platform and non-skid steps are much to be preferred. Where there are bridges over pipes on troughs on the floor, they should be given a non-skid surface. Ladders should be securely fixed and two persons should always be employed when long ladders are being used.

(i) *Air lines*. Air lines and hoses should be regularly inspected, water traps blown clear at intervals and filters renewed. If water from air lines is inadvertently blown on to electrical equipment, particularly printed circuit boards, a great deal of damage can be done. The practice of blowing dust off the clothing by means of an air line can be extremely dangerous and should on no account be allowed. It is all too easy for air to penetrate the skin, and air embolisms can prove fatal.

(j) *Fire-fighting equipment.* Fire-fighting equipment should always be accessible and all persons working in the moulding shop should know where it is and how to operate it. One of the best safeguards is a sprinkler system, the installation of which can save enough on insurance premiums to pay for itself in a very few years. However, the damage to material in the event of its operation can be considerable and it is wise to take extra precautions in its storage. Asbestos fire blankets are very useful, particularly if clothing catches alight or if the fire is small and can be easily confined. It is also a good plan to have a static water tank or fire buckets available at strategic points in the building.

(k) *Noise.* Noise in the moulding shop is a hazard to health and safety and should be prevented or reduced as much as possible. Sometimes panels are removed from machines during maintenance and are not properly replaced. Without them, the machine is not only more susceptible to the entry of dust and dirt, but it is also often noisier. With very noisy machines, thick blockboard screens can do much to reduce noise and to make for more pleasant working conditions. If portable granulators have to be used in the moulding shop, a strong blockboard cabinet surrounding each one will help to reduce noise and will also eliminate dust. Where an operator is using a large granulator for several hours at a time, ear muffs must be provided. Of course, the greatest single step towards the reduction of noise in the moulding shop (and to the better utilization of energy) will be taken when machines no longer have individual power sources, but are supplied from a central hydraulic fluid reservoir, with the motors and pumps in a separate sound-proof room. This used to be the situation in the days of compression moulding when many machines operated from a central supply system and the only noise was the gentle hiss of steam from a steam trap.

(l) *First-aid equipment.* Minor accidents are bound to happen and the provision of proper equipment for treatment may prevent them from developing into something worse. Trained first-aid personnel should be available on each shift and should be responsible for ensuring that the equipment is complete at all times. An eye-wash bottle is a useful additional item in case of an operator getting dust or other foreign matter in the eye.

Maintenance

It is foolish to wait until a machine breaks down before carrying out maintenance. It may also be unwise to do major maintenance work just because a planned maintenance schedule says that the date for doing it has arrived. However, Murphy's law of cussedness states that a breakdown will always occur at the most inconvenient time. If the breakdown prevents the execution of a large and profitable order, the loss incurred may be much greater than the cost of carrying out planned maintenance.

(a) *Daily checks.* Some maintenance items should be checked daily: i.e., hydraulic oil levels, lubrication, leaks on pressure lines, hose connections, and electrical wires, including thermocouple leads. Every week the machine should be cleaned down and the oil checked for cleanliness. A little less frequently, thermocouples should be checked by inserting the tip in boiling water and making sure that it indicates 100°C. Timers can occasionally be checked by a stop watch and the screw non-return valve checked for leakage.

(b) *Reporting faults.* Any reports of malfunctioning should be noted and the first convenient opportunity taken to put them right. Among the items that should receive regular, though not necessarily frequent, inspection are tie bars, tie bar bushes, carriage tracks and bushes, toggle link bushes, and lever pins. Hydraulic cylinders should be checked for leakage at the glands. If any undue wear is apparent, time may be saved at a crucial moment by taking the first available opportunity to put matters right. If this procedure is followed, it should not be necessary to carry out the planned maintenance exercise of stripping equipment and replacing worn parts at intervals of six or twelve months. However, a minute inspection of all working parts of the machine at three-montly intervals will generally pay dividends in ensuring steady production.

(c) *Spare parts.* Nothing is more frustrating than having to wait for parts to be sent from a distant supplier. The agent will often be able to recommend which items are most likely to fail and should be kept in stock. Often money is well spent in this respect as a great deal of time can be saved in waiting for spares. There is a good case for having several machines in the moulding shop from one supplier as this obviates the necessity for carrying too many spares. It is not always good practice, however, to 'tool up' with only one make of machine, particularly in the custom-moulding field. Not all machines have the same versatility and having only one make can be a limitation. However, it is very unwise to go to the other extreme and have a dozen different makes. Perhaps three different makes in a factory of 25 or 30 machines is a good compromise, particularly from the point of view of storing spare parts.

(d) *Ancillary equipment.* Planned maintenance concerns not only the injection moulding machines. All ancillary equipment, such as mould temperature controllers, granulators, stacking or conveying equipment, is also important. The main factory water-cooling installation should also be cleaned annually and de-scaled if necessary. In this respect, too, occasional de-scaling of mould coolant channels and cooling channels on the machine oil coolers can bring nothing but benefit.

Plant interior decoration and colour coding

It is always more satisfactory to work in good surroundings and even though the moulding shop may not be ideal from many points of view, there is no reason why it should not be made as pleasant as possible. Clean and brightly painted walls have many advantages, not the least of which is the conservation of light. An occasional coat of good quality emulsion paint has the dual effect of making the surroundings more pleasant and of helping to seal rough brickwork and so reduce dust. In any case, paint cannot be satisfactorily applied to dirty and dusty surfaces, so painting provides an opportunity for giving the factory a thorough cleaning.

(a) *Colour coding.* The colour coding of supply pipes is a useful expedient in helping to prevent the incorrect connection of services. There is a British Standard (BS 1710, 1975: Identification of pipelines) on the subject. In certain circumstances it could be damaging to a mould to circulate anti-freeze and, of course, air and water lines should not be confused. It may also be advantageous to have some easy means for recognizing different mains voltage outlets, apart from the different plugs and sockets. Lines can then be traced more easily to ancillary equipment.

(b) *Marking work areas.* It is good practice in the moulding shop to mark different areas of the floor for specific purposes. Gangways should always be free from obstruction and there should be a defined area by each machine for storing material for immediate use. Another area needs to be set aside for storage of finished goods before they are taken to the store or inspection department. Different coloured self-adhesive floor tapes are available for these purposes.

(c) *Using 'house' colours.* Perhaps to advocate painting all the machines in the factory 'house colours' might be considered a little unnecessary. There is some virtue in having a uniform colour for all the machines and no doubt, manufacturers would often be happy to oblige. Moulding shops where all the equipment is painted in this way certainly look very smart and this must generate some desire on the part of the personnel employed to keep the moulding shop clean and tidy. The creation of this attitude is a major step towards good housekeeping.

10

Keeping records

One of the jobs disliked most in the busy moulding shop is the keeping and maintaining of records. Some factories have very complicated systems necessitating a great deal of clerical work, and in very large shops the records may even be kept by computer. However, by far the largest number of organizations engaged in injection moulding in the United Kingdom are small, perhaps less than 100 having more than 20 or 30 machines. It is mainly for the smaller companies that this chapter is included, although many of the larger plants seem to have rather poor recording systems.

The question that first needs to be answered is, 'why keep records?' Having decided that some records are necessary, it must then be decided what *needs* to be recorded and what is *desirable* to have on record. These are two very different aspects, because it is *desirable* that records should be as complete as possible, but there is a constraint on doing this by the cost and time involved. Some compromise is therefore necessary.

Why keep records?

All injection moulding is carried out for one or the other of the following reasons: (a) To make a profit for the company or (b) for 'in house' moulding, to provide a service for the company in supplying mouldings or components as they are required at an economic price. To fulfil either function properly it is necessary to know certain facts about the business, some on a day-to-day basis, some at less frequent intervals, and some for reference at a later date to show how a job was done earlier, should a similar job arise. Day-to-day records are required in order to maintain maximum possible machine utilization and to ensure that adequate supplies of material are available to fulfil orders and so keep customers satisfied. These, together with records made at less frequent intervals, enable a check to be made of factory performance, costs, prices and profitability. File records are required mainly for reference, but can also be of use in legal disputes or when the time comes to re-equip the factory so that comparisons can be made of the performance of different machines or pieces of equipment.

Records, then, are desirable and necessary, but what are the basic requirements and what proportion of time should be devoted to them?

In the following sections of this chapter, six specimen forms for keeping records are outlined and discussed. They may not include all the detail that a particular organization requires; they may be considered to contain too little detail, or too much detail and as a consequence require too much time for completion. They have, at least, the merit of including most of the matters on which information is needed if a project is to be followed through the factory and completed satisfactorily. They will provide information from which machine utilization can be calculated, and on which parts of the organization are profitable and which are not. They will enable a check to be kept on material utilization and cost, and will help in planning to keep jobs moving. They will also provide a record in case of dispute or as a guide to future work. If filled in at the proper time they should not prove difficult to work with and will not use up valuable time and effort.

Forms 1 and 2 (Fig. 10.1 and 10.2) are filled in by the storekeeper who retains form 1 and, at the completion of a project returns form 2 to the office, where it is filed with form 4. Form 3 (Fig. 10.3) is clipped to the machine and is returned to the office each week. It should be filled in at each time of starting or stopping by the operator or setter. Form 4 (Fig. 10.4) is for the production manager's use and is retained in the office. Form 5 (Fig. 10.5) is kept on the machine for the duration of the job and form 6 (Fig. 10.6) is kept by the

STOCK RECORD

Material .... Supplier ....

Grade .... Stock Level ....

Colour .... Order Level ....

Date	Quantity received kg		Quantity issued kg		Job No.	Stock kg	
	Virgin	Re-ground	Virgin	Re-ground		Virgin	Re-ground

Fig. 10.1 Stock record

FINISHED GOODS RECORD

Job No. ....   Description ....

Customer ....

Date	Received from Moulding shop	Passed	Rejected	Despatched	Stock

End of project balance:
Material issued virgin ....kg
Material issued re-ground ....kg
Good mouldings despatched ....kg
Re-ground material in hand ....kg

Fig. 10.2 Finished goods record

MACHINE RECORD

Machine No. ....   Week ending (date) ....

Date	Job No.	Production	Time start	Time stop	Reason for stoppage

Fig. 10.3 Machine record

chargehand or foreman fitter until the mould is returned to store, when it is filed with the job record and the job report.

It will be seen that the job record is the most important document, because it brings together all the other information which is then filed in one place. For continuing jobs, a time limit can be placed on the use of this form and this can be made the occasion for a price review or a reappraisal of the project. This form is the basis of the system, but it is given fourth place in the discussion because, although a job starts by someone placing an order, its execution starts in the raw material store.

JOB RECORD (for office use)

Job No. .... Description ....

Date .... Customer ....

Material .... Possible machine ....

Weight (g) ....

Estimated cost/item ....(£) Estimated cycle (sec) ....

Mould commissioned (date) .... Mould No. ....

Mould maker ....

.... Promise date ....

Progress:

First trial (date) ....

Modifications:

Second trial (date) ....

Result:

Moulding started (date) .... Moulding finished (date) ....

Total time (hours) .... Total material (kg) ....

Total mouldings .... Total scrap (kg) ....

Total cost:	Material ....kg @ £	/kg	£
	Moulding ....hr @ £	/hr	£
	Extras		£
	Total mouldings ....@£	/000	£
	Job profit		£

Fig. 10.4 Job record

JOB REPORT (For use on machine)

Date .... Job report No. ....

Job No. .... Mould No. .... Machine No. ....

Material ....

Machine Conditions

Temperatures °C	Times (sec.)	Pressures (lb/in^2)
Cyl. 1......................	Inject	Inject ....
Cyl. 2......................	Hold	Hold ....
Cyl. 3......................	Screwback	Screwback ....
Cyl. 4......................	Cool	
Nozzle	Pause	
[1] Mould	Overall	
Oil		Screwback r.p.m. ....
Mould coolant		

Number of mouldings needed ....

Promises to customers:

Other information:

[1] There may be more than one temperature to record, e.g. for the various parts of the mould.

Fig. 10.5 Job report

Material balances

With the high price now charged for good quality plastic raw material there is a need to be able to account for virtually every kilogram used. This means that the basic raw material unit cost is about 40 p. Other materials, such as nylon, acetal, polycarbonate, etc., are three or four times that price and it is necessary with these materials to account for losses of anything over about 250 g.

The proverbial 'average moulding shop', with its two 8 oz machines, or equivalent, will be processing, on a 100 h week basis, 8–12 tonnes of material per week. The annual usage of, say, 500 tonnes would represent a value of at least £200 000, and considerably more if the more expensive materials were being used. If delivery averages four weeks, a raw material store of some 50 tonnes capacity is needed and, when dealing with these quantities and values,

MOULD RECORD

Job No.	Mould No. ....
Customer	Suitable for machine ....
....	....
Mould maker	....
....	
Date commissioned	Date received ....
Drawing No.	Where filed ....
Copy (ies)	Where filed ....
....	....
Drawing 1st modification (date)	Where filed ....
Drawing 2nd modification (date)	Where filed ....
Further drawings: Number	Where filed ....
Copies	....
Mould modifications (date)	Details ....
	....
Mould first trial (date)	Job report No. ....
Mould modification (date)	Details ....
	....
Further drawings: Number	Where filed ....
Copies	....
Mould second trial (date)	Job report No. ....
Mould repairs (date)	
Detail:	
Mould storage: Rack	Position ....
Date	Whether cleaned and greased Yes/No

Fig. 10.6 Mould record

a close check is essential. The loss of only 1 per cent of material over the year would incur a financial loss of £2000, enough to reduce profitability to an unacceptable level.

The first record required, then, is of material delivered: date, type, grade, colour, and quantity. Form 1, as an index card, one for each material variation, enables a stock balance to be struck after each transaction and provides the information from which material balances for each job may be obtained. Records of re-ground material can be kept on the same cards if desired.

In the small company the raw materials storekeeper may also act as the despatch clerk. Form 2, to be used for finished mouldings, records the total numbers received from the moulding shop for each job, those passing inspection, those rejected, and those despatched. *All* mouldings received, however poor the quality, should be recorded, if not by numbers, at any rate by weight. Many moulding shops make the very grave mistake of not having a complete material balance, which this form enables the storekeeper to draw up. They have no idea how efficient or otherwise the processes are, and as a consequence jobs get wrongly priced and business may either be lost or executed at uneconomic rates. Form 2 can be used either for keeping a daily balance or for a complete job balance.

The importance of maintaining a record of material used and of comparing this with actual goods produced, cannot be too strongly emphasized. All too frequently, small companies find themselves in financial difficulties because of neglect of this point. Machines operate for several hours, perhaps, producing mouldings that have to be rejected, or multi-impression moulds produce less than their full quota of mouldings at each shot. A *laisser faire* attitude is adopted because all the waste is re-ground and used again. It must be realized that the raw material used then costs the original price, plus the cost of processing it, plus the cost of re-grinding, of sorting out good mouldings from bad and, with some faults, having completely to scrap some material. This is the situation in which keeping a record of material balance is most valuable, particularly in conjunction with machine operation records.

Machine operation records

Most factories have a budgeted machine utilization figure, based on the total number of hours of operation of the factory per week and the number of hours during that time which it is expected the machines will be in production to give a satisfactory level of profitability. Thus, for a 5-day week of three shifts per day, 8 h per shift, and total weekly hours of 120, the budgeted machine utilization might be 100. Taking the total costs (less raw material) of operating the factory and dividing by the number of machines and the number of hours (100) gives the machine hourly rate. Thus, if the machine hourly rate is £5/h, the income from the machine at 100 per cent budgeted utilization would be £500. A utilization of 100 per cent is very rarely reached (on the figures given above, it will be realized that 120 per cent is possible!) and many factories work at between 60 and 90 per cent.

To arrive at utilization figures for individual machines a simple record needs to be kept of machine productive time, that is, the total time per shift, day, or week that the machine is actually operating and producing mouldings. The record can be kept, if preferred, by just noting all the 'down time'—time when the machine is not functioning—and subtracting it from the budgeted time. Form 3 illustrates a suitable means of making the record.

The reasons why the machine is not working should be noted and might include mould changes, breakdown, waiting for material, mould repairs, or shortage of suitable work. This kind of information has both ephemeral value and long-term importance. The feedback to management can, for example, help in the urging of supplies or spare parts; salesmen might be motivated to find more work for the machine, and a machine having a bad breakdown record might be considered to be due for redundancy. The record also enables responsibility for non-operation to be established and for action and preventive measures to be taken.

Job records

Form 4, the job record, is the key to the whole of this suggested system. As soon as an enquiry is received by the sales organization it should be allocated a job number. This number remains throughout its progress, even though the job becomes moribund. It is filed under the number, with suitable cross-referencing to the firm involved and the type of job. On the form are noted all the relevant details, including commissioning of moulds, the progress of the mould, results of trials, and estimating information; when the job is finished (or with continuing jobs, at prescribed intervals), a summary of the material used, numbers of good and bad mouldings made, and job profitability should be recorded.

As soon as the job goes to the moulding shop, a job report is started. This is for use on the machine and gives details of all conditions of operation. It also has a record of the total number of mouldings required, any promises made to the customer, and other important information that may be necessary. By having these details in front of the operator or setter, they are made aware of any special conditions or any other matters of urgency (form 5).

These records are relevant to the whole running of the factory. They enable the supervisor to arrange the work-load, to plan for mould changes, maintenance periods, etc., and generally to work out priorities.

Mould records

Almost as important as the job records are the records relating to the mould (form 6). Each mould has its own record because a job might have several moulds. Each mould record has a reference to the job number. The mould record starts when the mould is commissioned. Information about any modifications required or agreed should be kept with the record, as should information about maintenance, trials, etc.; and a note should be made of where drawings are filed. The mould should be given a number, indelibly stamped on it, and this number is also used on the record form.

The importance of keeping mould records hardly needs stressing. Moulds are generally quite costly and no one likes paying for them. Customers may not appreciate the cost of modifications and, though verbally accepting and agreeing to pay, may argue when the bill has to be paid. All changes involving any cost or charges whatsoever *should be agreed in writing* and copies of the letters filed with the mould record. It is better to do this than to face a legal battle later.

Another reason for keeping adequate mould records is so that, when setting up the mould, all auxiliary equipment can be connected without delay. All the relevant information will be to hand, because drawings can easily be found and, even though the mould may not have been properly labelled, there will be no likelihood of mistakes being made.

By the use of forms such as have been suggested here, a great deal of information can be collected which will be of help in carrying out the other tasks of management. Estimating will become easier as more information on factory performance is made available. Progressing work and establishing a factory programme becomes more positive; but perhaps the greatest benefit to be derived from the keeping of a few simple records, such as those illustrated, is summed up in the words of the old Chinese proverb: 'The strongest memory is weaker than the palest ink'.

11

Factory layout, services and ancillary equipment

The proper layout of the factory and the provision of suitable services and equipment are very important factors in the efficient operation of injection moulding. It is very often found, particularly with small companies, that the premises are by no means ideal and that, even if they are quite reasonable, the best use is not made of them. Frequently, premises become overcrowded, the requirements for machines exceeding the available space. Under these conditions, efficient working is very difficult, machine maintenance becomes impossible unless carried out by a contortionist, and housekeeping and cleanliness go by the board. In this chapter, which is concerned with recommendations for efficiency, the ideal layout and services will be considered, and the main requirements of the small-to-medium sized factory will be elaborated.

Let us assume that a new injection moulding shop is being planned and that a suitable building is to be equipped. The first considerations are services—water, electricity, compressed air, and drainage. There may also be a call for chilled coolant for some moulds and it may be necessary to instal automatic handling equipment and conveyor belts. Provision will need to be made for raw materials and finished products stores, mould storage and handling equipment, storage of spares, and for the parking of trucks and lifting gear. There will be ancillary equipment to consider, such as mould temperature controllers, granulators, and mixing equipment. Space heating and lighting, and the provision of staff amenities such as cloakrooms, toilets, rest rooms, canteens and offices, are really the concern of the architect, but must not be overlooked at the planning stage. All these matters are relevant to the efficient running of the moulding shop and need to be considered in this context.

Power requirements

The first matter to decide is the amount of electric power required. This will normally be the sum of the requirements of all the machines, factory and office lighting and heating, ancillary equipment, ventilators, and office machines. When all this has been totalled, bearing in mind that some will be three-phase, 440 V, and some single-phase, 240 V, the total will be in excess of what will normally be required. Not all equipment is likely to be drawing its maximum rating at any one time and about 80 per cent of the calculated total requirements will probably be adequate. However, allowance must always be made for any expected (or possible) expansion because replacing the electric power supply company's cable can be very expensive.

Having arrived at a figure for the power requirement, terms then need to be agreed with the supply company if a heavier duty cable or additional sub-station are required.

Power factor correction

Two types of loading apply to the power supply to the factory: the resistance loading of heaters, lighting, etc., and the inductive loading of the motors. These need to be separately calculated and the sum of the total kilowatts of motor capacity, not forgetting pumps, compressors, granulators, etc., is of importance. With induction motors fitted to most machines and pumps, the relationship of power in kilowatts (kW) to kilovoltamperes (kVA) is generally low. This ratio, kW/kVA, is the power factor and if it is too low, the overall cost of electric power becomes high because the company charge for power is based on both kW and kVA. In order to improve the power factor, capacitors are installed and the system, called power factor correction, may increase the ratio from say, 0·7 to 0·95. There is little point in trying to achieve a power factor of unity because of the additional cost of equipment. With variable loads, the equipment operates by switching in or out different sizes of capacitors as the load varies.

Power factor correction is well worthwhile to instal. The equipment can be purchased outright or it can be leased or hired. On the basis that an account for £1000 incurred with a power factor of 0·7 would be reduced to about £750 if the power factor were increased to 0·95, the cost saving is considerable.

Disposition of machines

With the power requirements settled, the next phase is to decide on the services and the positions of the various service points. For this, it is necessary to know the positions of the machines and this depends on their size and the size and shape of the building.

In siting machines of up to about 500 tonnes lock it is desirable to allow up to 2 m between them, but if space is limited, it is possible to work in a clear space of 1·5 m. The best plan is to find from the machinery specification the actual machine dimensions and then to leave sufficient space between them to take in the gantry or fork-lift truck required for mould mounting. It is also

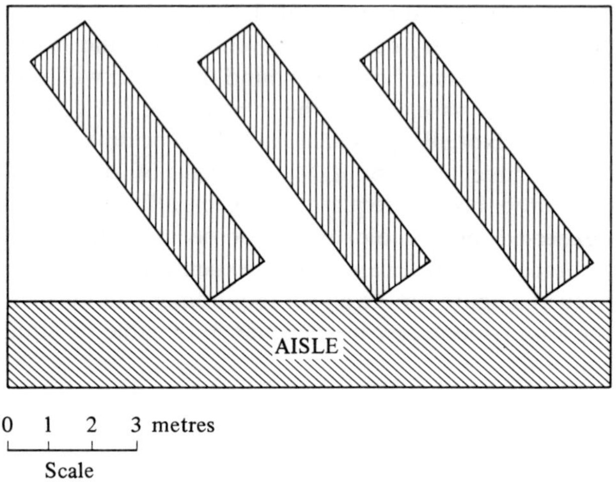

Fig. 11.1 Machine arrangement—echelon

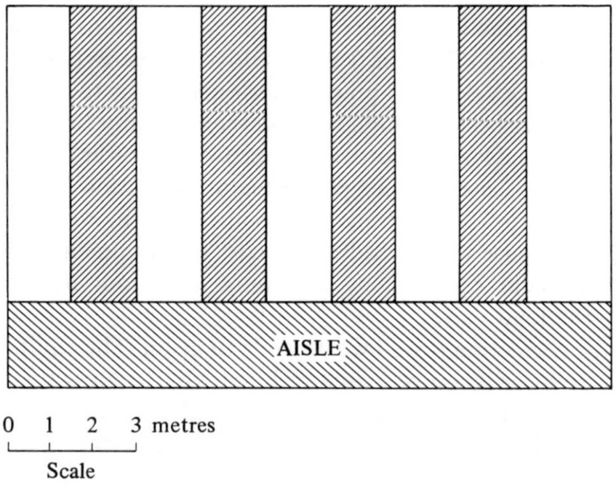

Fig. 11.2 Machine arrangement—parallel

desirable to leave sufficient space for a mould temperature controller and for a set of steps, if necessary, for reaching the feed hopper.

The disposition of machines in relation to each other is a matter for argument. Some moulders favour the idea of installing machines back-to-back in groups of four, on the basis that one operator can look after four machines. Others favour the echelon formation, illustrated in Fig. 11.1, but this takes up more space than the formation shown in Fig. 11.2. Whichever arrangement is decided upon, this will determine the positions of electrical connections, drains and water supply points. The question as to whether automatic stacking or conveying equipment will be required has to be decided upon at this stage.

Service lines

There are two main ways in which services can be carried to the machines: overhead or underfloor. In general, the former is preferred. If open trenches are used there is always the possibility of flooding, with its attendant danger to electrical services. Dust and dirt can easily penetrate an open drain and foul the cooling water. Overhead services also have disadvantages. They prevent the free movement of gantries and A-frames and can sometimes be more costly to instal. On balance, however, the advantages outweigh the disadvantages.

Both water and electrical services are best served with ring mains because it is possible to use smaller pipes and smaller gauge electrical conductors without overload or reduction of pressure and voltage. Connection points for all services should be taken as near as possible to the machines, but to reduce the possibility of vibration causing fracture of water pipes, the final connection to the machine can be by means of a flexible hose. Electrical connections from overhead busbar can be through conduit pipe and flexible tube direct into the control cabinet of the machine. In a factory with two rows of machines, alternative arrangements for layout and services are suggested in Fig. 11.3. If the machines are along the walls of the factory, with a centre aisle, pipework and busbar can be clipped to the walls. Using the centre service line requires gangways around the perimeter of the building, which may be advantageous from the point of view of safety. In this case, the overhead services would be clipped to the roof members, with connections brought down to a suitable height.

Air lines, refrigerated coolant pipes, and auxiliary services, such as 240 V single-phase outlets and sockets for 440 V three-phase connections, can be sited to suit requirements. The former can be clipped to the same supports as the water services, the latter are quite convenient if fixed to the machine control cabinets. It is also convenient to have some outlets of each type on the walls, at suitable positions, bearing in mind that machines such as granulators, mould chillers, and some mould temperature controllers require a three-phase supply. Most other ancillary equipment, such as mould heaters, power drills, and smaller mould temperature controllers use a single-phase 240 V supply.

When considering the total electrical requirements, the lighting of the factory is important and uses a significant amount of power. Standards are recommended for lighting intensity and should be adhered to. Ill-lit factories can be a cause of accidents and can also be detrimental to the sight of workers. Alternate rows of fluorescent lights are generally best wired into different phases so as to reduce any stroboscopic effect where there are moving parts.

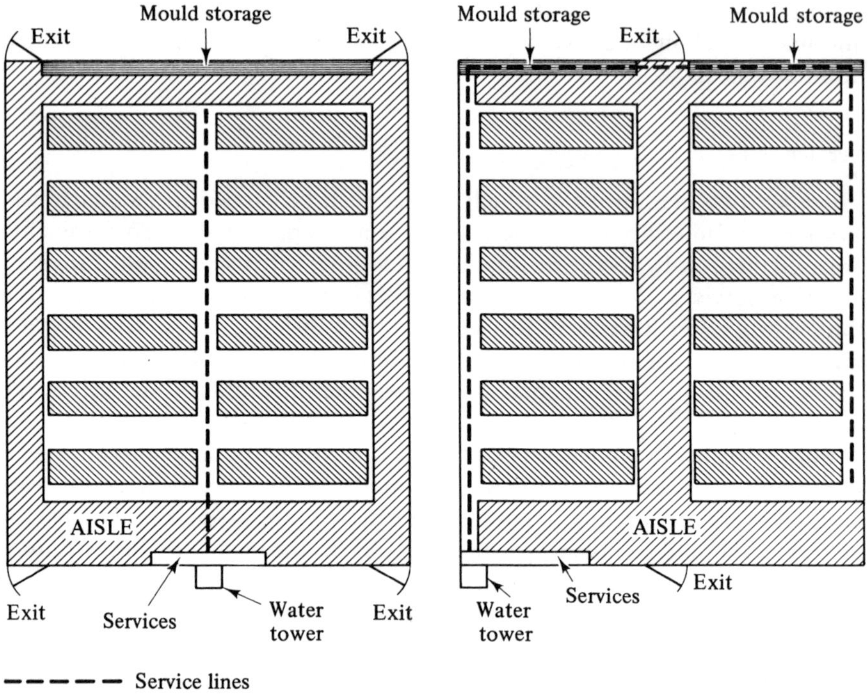

Fig. 11.3 Factory layout and services

Cooling water supply

The provision of adequate supplies of cooling water throughout the factory is of the utmost importance. Some machinery manufacturers make recommendations about the quantity of cooling water their equipment requires and will sometimes quote figures for different inlet temperatures. Table 11.1 gives some typical figures for inlet temperatures of 15°, 20°, and 30°C.

Not all manufacturers commit themselves in this way and it then becomes necessary to make some approximations. It has been found by observation that, using fixed-delivery vane pumps as a source of hydraulic pressure, about

Table 11.1 Cooling water requirements for machines (inlet water temperature)

Machine size tonnes lock	Gallons/hour			Litres/minute		
	15°C	20°C	30°C	15°C	20°C	30°C
30	40	70	140	3	5·25	10·5
50	50	80	160	3·75	6	12
75	55	88	180	4·1	6·6	13·5
100	65	100	200	4·9	7·5	15
175	70	110	220	5·25	8·25	16·5
220	77	130	260	5·8	9·75	19·5
300	90	140	280	6·75	10·5	21

one-half of the energy put into the system as electricity has to be taken out as heat. In a typical installation of 16 machines, the total power usage was 194·5 kW or 260 hp. Taking 1 hp = 2500 Btu/h or 630 kg cal/h, this installation would need cooling of:

$$2500 \times 130 \text{ or } 325\,000 \text{ Btu/h or } 630 \times 130 \text{ or } 81\,900 \text{ kg cal/h.}$$

Assuming a temperature rise of just over 5°F or 3°C, this would require a flow of about 6000 gal/h or 450 l/min. Obviously, to take this quantity of water from the town mains (or even a quarter of it, allowing a temperature rise of 22°F or 12°C) would be extremely expensive and wasteful and a cooling system was required.

A typical cooling tower is illustrated in Fig. 11.4. In the simplest system, warm water from the factory is returned to the top of the tower and cascades down through the tower packing, meeting an upward current of air drawn through the tower by the fan in the chimney section. Here it is cooled, chiefly by evaporation, and collects in a reservoir in the base, from where it is pumped to circulate around the factory.

A tower for extracting heat at the rate of 325 000 Btu/h would probably have a reservoir of about 200 gal with 50 gal in the pipes, a total of 250 gal or 1125 l. If all the cooling is provided by evaporation it will be necessary to evaporate 32·5 gal (140 l) of water per hour (latent heat of vaporization of water at 30°C is about 1000 Btu/lb or 580 kcal/kg). In about eight hours the whole of the water in the system would be evaporated, but the quantity is continuously made up from the mains. This is, of course, much less than would be required for direct cooling and shows a considerable cost saving.

In many parts of the world, town water supplies are 'hard', that is, they contain quantities of chalk and magnesia in solution. In the United Kingdom, notably in south-east England, south of a line from The Wash to the Bristol Channel, East Yorkshire and parts of the North-West Midlands, the mains supply contains from 50 to 300 parts per million (0·005–0·03 per cent) of calcium and magnesium salts. If this water is used in a simple evaporative tower, the concentration of the salts would increase by a factor of ten in less than a week. At this stage, precipitation of the chalky material occurs in the hottest part of the system and cooling channels rapidly scale up. Water softening is sometimes used to reduce this effect, but the normal system used, ion exchange, simply substitutes sodium for calcium and magnesium and then the water becomes progressively more alkaline and somewhat aggressive to the softer steels.

Even in soft water areas, such as South Wales and Scotland, and especially in urban districts, the water in an open tower system is likely to become contaminated with atmospheric pollution, such as soot, sulphur dioxide and, in country districts, dead leaves, etc. The water also becomes highly oxygenated, making it possible for rust to form more easily. All these matters are detrimental to the efficient running of the moulding shop and can cause problems in the quality of the moulded components.

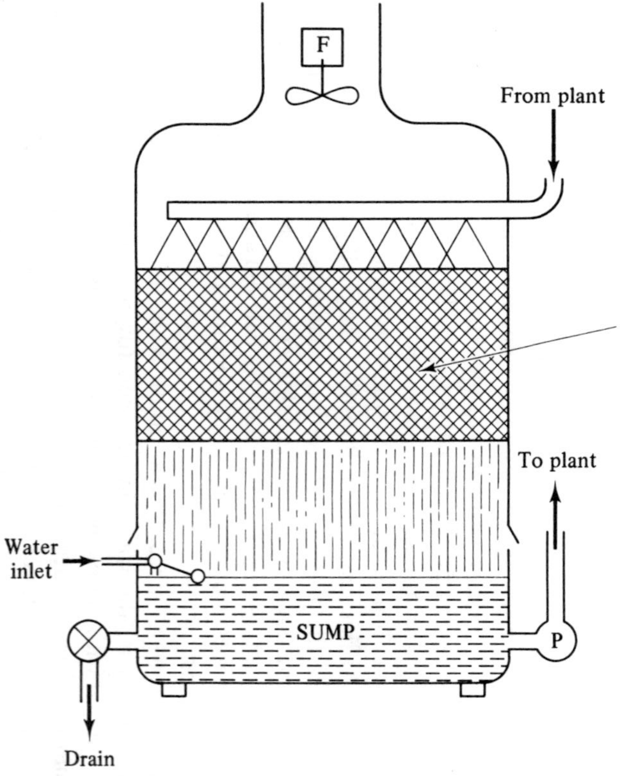

Fig. 11.4 Cooling tower—direct

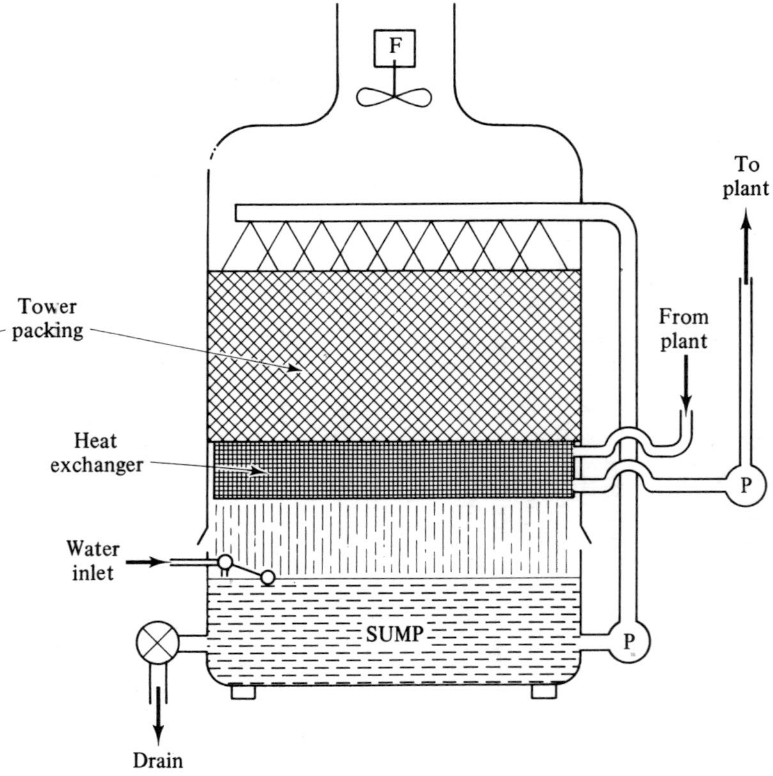

Fig. 11.5 Cooling tower—indirect

Additives are reputed to overcome many of the problems and will keep the water clean. They are satisfactory with low levels of pollution and hardness, but in hard water areas the cost is likely to be prohibitive. In such areas an indirect cooling tower is preferred.

Figure 11.5 illustrates the principle of the indirect cooling tower, using a heat exchanger. Instead of being circulated to the top of the tower, the warm water from the factory flows through a heat exchanger and then into a balancing tank before being pumped back into the factory. Cooling of the heat exchanger is, as with the open system, by evaporation of water pumped from a sump in the base of the tower to cascade through the tower packing and then over the fins of the heat exchanger. Obviously, the same amount of evaporation occurs as with the open tower, but because the maximum temperature is rarely in excess of 30°C the chalky material is precipitated as a soft sludge that collects in the bottom of the sump tank. Periodically this can be removed by draining the sump and if any scaling of the outside of the heat exchanger occurs, this is easily removed by means of a dilute organic acid. The water in the system supplying the machines and moulds is not evaporated; it stays at the same concentration of calcium and magnesium salts and progressively becomes de-oxygenated and has less tendency to form rust or scale. In some installations, it has been found possible even to drink the water after it has been circulating for a year.

Indirect towers are not quite as efficient as direct ones and are considerably more expensive. In hard water areas they can save the extra cost in about one year's running by increasing efficiency, less clearing of cooling channels, and steadier running. All types of cooling towers tend to give some variation of cooling water temperature according to the state of the weather. On a hot, humid day in summer, the efficiency is lower and the temperature of the cooled water may rise to 5°C above ambient, but under most conditions they will cool the water to about 2·6°C (5°F) above ambient. Towers should be sited out of doors in the coolest part of the factory precincts, if possible by a north-facing wall. In very cold weather the pipework and sump tank should be fitted with thermostatic frost protection devices.

Some moulders use the heat from the machines and moulds for heating the factory. This has many advantages and provides a way in which the water can be cooled without causing contamination. The heat is, however, 'low-grade heat' and, unless a heat pump system is used, can provide only low-temperature heating. Because changes in the weather can bring about changes in the demand for heating almost from hour to hour, this kind of system needs to make the appropriate provision and could prove expensive.

Various types of service pipes can be used and each has advantages and drawbacks. Galvanized iron is very satisfactory and with the pipes being rigid and self-supporting, they only need to be clipped to supports at quite considerable spans. Mild steel or soft iron tend to rust easily, both inside and out. Copper is too expensive and, although relatively cheaper, PVC or polythene pipes need frequent supports and are prone to sagging. Allowance also has to be made for the high thermal expansion of the plastic. Galvanized iron of 2–3 in diameter keeps its appearance without being painted, and pipes of that diameter, operating at a pressure of 40–80 lb/in^2 (3–6 kg/cm^2) will allow a flow well in excess of 10 000 gal/h. For the main circuit, a pump of 5–10 hp is required, depending on the pressure and flow, and for the tower pump, about 4 hp.

With totally enclosed systems where there is no visual indication of flow, it is desirable to fit either flow meters or flow indicators which move when the water is turned on. Thermometers at strategic points are also useful in indicating the effectiveness of the cooling, although for the hydraulic system the best indication is the oil temperature gauge. The great advantage of the totally enclosed system, such as has been discussed, is that there is generally no need to reduce the flow of water by regulating a tap. The full flow through the system will normally maintain the temperature within the desired limits.

Mould temperature controllers and chillers

It is not always desirable to cool moulds from the general supply, because it may not be running at the ideal temperature for the material being moulded. Neither is it good practice to throttle back the supply of coolant through the mould in order to achieve the required temperature, because of the danger of getting hot spots in the mould. Less than maximum cooling efficiency also results from this practice. Mould temperature controllers working in conjunction with the main cooling system provide the best solution.

Two main types of mould temperature controller are available: the simplest system has a tank, connected through a ball valve, to the water supply. In the tank is a thermostatted water heater and a copper cooling coil connected on the inlet side to the water supply and on the outlet side through a solenoid valve (controlled by a thermostat) to the drain or return pipe of the cooling system. The water from the tank is circulated to the mould, using a small pump. If the water becomes too hot, the valve on the cooling circuit opens and the water in the tank cools to the required level. A similar arrangement can be adapted for oil or glycol circulation, but in this case the tank is not filled from the mains.

In the other system, water from the mains passes into a heating chamber and is circulated by a pump to the mould. If the water becomes too hot, a solenoid valve on the outlet side allows the hot water to flow out to the drain and cold water flows in from the mains to take its place.

For some moulding jobs, mould chilling is employed and chiller units having refrigerated heat exchangers are used. They are fairly expensive and costly to run and, if mould cooling channels have been properly designed and are adequate in size, they should not be necessary. The pipework connecting chiller units to moulds should be insulated because the temperature of the coolant is likely to be well below the dew-point of the atmosphere and will cause condensation. Glycol or methanol/glycol anti-freeze solutions are added if the temperature is to be allowed to fall below the freezing point of water.

As discussed in Chapter 5, the size of mould temperature controller is of importance. Knowing the shot weight, the cycle time, and the enthalpy of the plastic material, flow of coolant required can be estimated by assuming that half the total heat of the material will be removed in the mould. Sufficient coolant flow to allow a temperature rise of not more than 5°C or 9°F will then be needed.

Granulating and mixing equipment

Most injection moulding organizations cannot go to the expense of having a complete reprocessing plant consisting of granulator, mixer/extruder, and die-face or lace cutter. The general arrangement is to have a room in a fairly remote corner of the building where material can be granulated using an ordinary masson-type cutter. Although this produces re-work of irregular-shaped particle size, with a fair amount of powdery material, it is perfectly satisfactory for injection moulding.

(a) *Granulators*. The normal granulator has a powerful electric motor driving a heavy rotor, to which are clamped two or more stout cutting blades. The rotor and blades are a clearance fit in a cylindrical chamber having a set of stationary knives. A long feed throat with spring-loaded baffles prevents injury to the operator and also stops pieces of material from flying back from the cutting chamber. About one-third of the cylinder wall of the cutting chamber consists of a screen of suitable size to allow granules for moulding to pass through and a tray underneath catches the granulated plastic.

This type of machine is extremely noisy and the operator wears ear muffs. It also creates a lot of dust and must be against an outside wall, and well ventilated. The walls of the room or cabinet should be of double skin or of sound-insulating block board. One important hazard which must not be overlooked is that of dust. Any accumulation of plastic powder should be prevented because, if this were to be disturbed and form into a dust cloud, a spark—even static electricity—could cause an explosion. Material for granulating must be kept free from dirt and moisture and when the re-grind is being used in the moulding machine, a magnetic screen to prevent access to the cylinder of any small pieces of ferrous metal from the cutting knives should be inserted in the feed hopper.

(b) *Mixing equipment*. Mixing equipment can be simple or complicated. For small quantities of up to 5 kg, a polythene bag of double the capacity of the material can be used. The ingredients are placed in the bag, which is then screwed up from the open end to seal it and to trap a volume of air. This allows the material to be shaken about in the bag and gives extremely good mixing in a few minutes. For larger quantities, mechanical mixing is desirable.

The simplest mechanical mixer consists of a drum capable of holding, say, 50, 100 or 200 kg of material. This is mounted on a framework which can be rotated by an electric motor, causing the drum to tip end-over-end. Not more than half the capacity of the drum should be mixed at one time. Some equipment is so arranged that the drum does not rotate end-to-end but diagonally, corner to corner. Baffles may also be mounted inside the drum. Many machines are designed to accommodate the standard drums used in industry.

More sophisticated mixers work on the same principles but have variously shaped mixing chambers, with a view to getting better mixing action. Thorough mixing takes time and it cannot be expected that the best blending, say, of colour in a 100 kg batch would be achieved in less than 20–30 min.

Where large quantities of material, such as coloured masterbatch and natural polymer, are to be mixed, mechanical equipment which meters the correct amount of each ingredient from separate bulk storage silos is used. The material passes along a tube having in it an Archimedean screw which rotates and gives a mixing action as it conveys the plastic. Other systems use a tube with a series of baffles which cause turbulence in a current of air used to convey the plastic and so bring about mixing. It should be remembered that any device, such as a mixer or conveyor, that introduces a current of air into the plastic, whether directly or by breaking a vacuum, should be looked on as a means of allowing the ingress of moisture. Such devices would not be recommended for use with the more hydrophylic plastics, such as nylon, acrylic, ABS, and polycarbonate, unless the air was completely dry.

Hopper loaders

Where large quantities of material are being processed, some means of loading the hopper, other than manual, is necessary. If the material is supplied in bulk containers which can be hoisted to a mezzanine floor, the hoppers can be directly fed by gravity. Where this is not possible, there are many systems available on the market for loading the hopper either from bulk storage or from 25 kg bags.

It is not the function of this section to adjudicate on the relative merits of the various systems, but simply to point out where problems can arise. As with mixing systems, so with conveyors and automatic loaders, care must be taken to prevent material picking up moisture. It is most likely to do so if undried air is used or if a vacuum system is operated. If air can be dried before coming into contact with the plastic, so much the better. If the plastic is *at a lower temperature* than the air, it will pick up moisture. Let it be understood that, when something is 'dried', it simply means that moisture has been taken from the material and put somewhere else. Air is the commonest conveyor of moisture, and care must be taken to see that it conveys it *away from* the plastic, and not *into* it.

Hopper driers

These rather expensive pieces of equipment are designed to make sure that material is dry before being fed to the cylinder. They can, however, present a problem. They operate by heating a current of air which flows through the plastic as it resides in the hopper, and therefore for any given size of hopper, there is a given rate at which material can be dried. As drying takes several hours, except for very small machines, the hopper would need to be very large. There is a danger that, on the assumption that the hopper drier is going to do its job, the machine operates, in fact, more quickly than the material is dried.

Hopper driers are best used in conjunction with a conventional drier and will serve their best purpose by 'putting the finishing touches' on the drying operation by keeping the dried material warm and dry whilst it resides in the hopper.

General drying equipment

There are two general methods of drying plastics:

(1) By blowing a current of hot dry air over the material until it is hot and gives up its moisture.
(2) By heating the plastic, removing all the surrounding air, and causing the moisture to evaporate under reduced pressure (vacuum drying).

In method (1) it is necessary to allow sufficient time and to make sure that the air is as dry as possible. There is often a limit to the temperature to which the plastic can be raised, in general, something less than 100°C. Prolonged exposure to air tends to oxidize some materials. Table 11.2 indicates the optimum drying temperatures, with the maximum allowable moisture content for good moulding, for a number of plastics.

Two types of hot air drier are available. One consists of a drum having three, six, or more perforated cans which fit in the upper part. In the base is a

Table 11.2 Drying temperatures and maximum permissible moisture content for good moulding

Material	Drying temperature (°C)	Moisture content (%)
ABS	80	nil
Acrylic	75	0·05
Acetal	Should not need drying	
Nylon type 6 and 66	85–110	0·25
Nylon type 10 and 12	80–100	0·06–0·10
Cellulosics	75	0·40
Polyethylene	75	0·07[1]
Polypropylene	80	0·10[1]
Polystyrene	70	0·10
Polycarbonate	120	0·02
PVC	65	0·08
SAN	80–90	0·10

[1] These materials do not normally need to be dried, but certain grades containing carbon black may have a tendency to absorb a little moisture.

small fan which blows air through electrical heaters situated below a perforated false bottom on which the canisters rest. A lid at the top allows the air to escape. The cannisters can all be filled with the same plastic or with different materials, because there is not enough draught to distribute any dust or to cause contamination.

The other type consists of an oven with shelves on which are placed shallow trays containing thin layers of the material to be dried. Nylon gauze is used to cover the trays to prevent the spread of dust particles. Air is blown over the shelves from a heater and fan at the side. In both types of drier, the air temperature can be thermostatically controlled, fine gauze over the inlet filters out any dirt and dust, and the normal time for drying, dependent upon the amount of moisture in the material, is 2–6 hours.

In method (2) the plastic is placed on trays in an oven which can be evacuated. The shelves on which the trays are placed are generally heated to a controlled temperature, and heating must be by radiation or conduction because there is no air present to act as a heat-transfer medium. This type of drier is very satisfactory, provided that the necessary temperature for proper drying can be achieved without charring or melting the plastic. When the vacuum is broken, prior to opening the oven, dry air must be admitted, otherwise all the benefit of drying can be destroyed very quickly.

Raw material and finished product stores

The cost of storing raw material and finished goods is not inconsiderable. In a factory using 1000 tonnes of material per annum it would be prudent in normal times to carry about 100 tonnes of stock, that is, to allow for delivery times of four or five weeks. With large usage it is often possible to arrange for weekly deliveries, but this would probably be only for one type and colour or for natural material. Stores have to be accessible by fork-lift trucks, which must be able to operate without damaging or tearing other sacks or consignments. It is desirable to have a heated store, although a temperature a little lower than the factory temperature can be tolerated. It is undesirable to stack material out of doors or in unheated sheds, particularly in cold weather, because of the danger of picking up moisture by condensation if taken into a warm factory and opened before reaching ambient temperature. Keeping too much stock is unwise (unless inflation is rampant and will pay for interest charges as well as storage costs), but it is equally foolish to reduce stock levels so much that machines are kept idle for lack of material.

Finished product stores should be adequate to store the amount of mouldings that constitute a delivery load. Space must be set aside so that there is no congestion due to normal accumulation of components. Many articles, for example, boxes, take up a great deal more room than the plastic raw material from which they were moulded, and the moulding shop can easily become overcrowded after a busy night shift.

It is a moot point whether packing material should be thought of as raw material or whether it should be held in the finished product store. Certainly it needs to be controlled by the storekeeper so that any wastage or damage to crates, cartons, etc., can be minimized.

Treatment and care of re-work

Mention has been made in several places about the need to take great care of re-work. This is particularly true of moisture-sensitive materials, but with the high cost of all plastics, most companies can ill afford to waste any material at all. Contamination of any kind must be excluded and, for this reason, scrap mouldings should be carefully handled, kept clean on the machine, and stored in closed sacks. Many small companies use old raw materials sacks for this purpose, and this is quite in order if the sacks are clean, can be closed, and contained the same material as the scrap consists of. A few foreign particles can contaminate a whole sack of re-work. If a variety of materials is being processed in the moulding shop, different kinds of scrap should be adequately labelled and segregated by type and colour in different sections of the store.

Regranulating can be an objectionable chore in the small shop. No one really wants to do it and the work goes by default until formidable quantities of scrap have built up. If it is attended to each day, the task will not become too great and the material balance will be much easier to maintain.

Table 11.3 Percentage of re-worked material passing through the machine one to seven times with 25, 50 and 75 per cent re-work

Number of times passed through machine	Percentage passed through machine		
	25% re-work	50% re-work	75% re-work
1	25	50	75
2	6·25	25	56·25
3	1·56	12·5	42·9
4	0·39	6·25	31·64
5	0·10	3·13	23·73
6		1·56	17·8
7			13·35

The question is often asked, 'how much re-work can be used?'. This depends entirely on the material and the job. There is no doubt that every pass through the machine is detrimental to the material, but deciding when the limit is reached is generally a matter for experimentation. Table 11.3 indicates the amount of material with one to seven passes through the machine using 25, 50 and 75 per cent re-work.

Perhaps the best advice to give is: 'Make the smallest amount of scrap possible with the minimum of sprues and runners, so that regranulating and the use of re-work does not become a problem'. One of the purposes of this book is to indicate some of the ways in which this can be done and to help the moulder to reach that happy state of affairs.

12

Financial aspects

Injection moulding is a capital intensive operation and is most profitable when carried out on a continuous basis. The vast majority of products are capable of being made fully automatically without an operator, or at most with a single operator or supervisor in charge of a number of machines. Arrangements can be made for automatic replenishment of the machine feed hopper from a central reservoir of material and the moulded components may be carried away on a conveyor belt or transferred, as they are made, to a stacker or to packages for despatch. Such fully automatic operations are obviously expensive to instal.

During the first quarter of 1976, machines of 1 oz (30 g) shot capacity having a 45 tonne lock were costing between £8000 and £15 000, depending on the country of origin and the degree of sophistication of the control equipment. Moulds similarly varied greatly in price, but for a machine of the foregoing size could cost up to £3000. Ancillary equipment was also expensive, and for providing a fully automatic process the total capital cost could be increased by another 60–70 per cent. Since that time prices have increased by up to 40 per cent.

The period of inflation between 1974 and 1978 saw a corresponding rise in the price of plastics materials. Being largely manufactured from crude petroleum, the large increase in world prices of that commodity caused a similar change in the price of plastics. Whereas in 1972 large quantities of the commonest materials—polystyrene, polyethylene, and polypropylene—were available at very low prices, by the beginning of 1975 prices had doubled or even trebled and most of the large suppliers have had to increase prices by at least 10 per cent per annum since that time. Putting this into 'shop floor' terms, a 25 kg paper sack of the cheapest plastic is now worth more than £10.

It is against this background that the plastics industry, in general, and the injection moulding part of the industry, in particular, are maintaining an annual growth rate of about 10 per cent worldwide and in some markets are growing even more rapidly. It is in these market conditions, too, that great attention needs to be paid to the financial aspects of injection moulding, since a small percentage loss caused by incorrect costing, poor estimating, or inefficient working on the shop floor, can rapidly lead to a position of financial insolvency. Any savings effected in materials costs, shorter operating cycles, reduction in power requirements, or lower labour costs, are likely to have a large impact on the profitability of the process.

Fixed and variable costs

The main fixed cost of the injection moulding process is, of course, the capital cost of the equipment, but other fixed costs—rent of property, rates, etc.—must not be lost sight of. In the world of the entrepreneur—and many companies in the field of injection moulding were started by such people—machine capital cost has not seemed so important in the past. Starting operations with one or two small secondhand machines, the minimum of ancillary equipment, and often poor or even inadequate premises, it has been possible by meeting a local need, perhaps in a specialized field, to make a very reasonable profit. This may have been achieved by hard work and vigilance on the part of the owner of the company, but it has enabled many small companies to grow and invest in newer and better equipment in the knowledge that any new capital expenditure has been met out of profit. It is probably fair to say that, of some 1500 companies in the United Kingdom engaged in injection moulding, perhaps half of them have operated, and many still do operate, in this way.

The owner-manager of such a company does not need to consider the servicing of his capital at, say, 15 per cent per annum. He only considers the loss of interest, less tax, on his capital which may, ultimately, be no more than 5 or 6 per cent. In such a situation, quoted prices tend to be lower than the general market prices. The owner-manager can afford to be selective in his business, avoiding projects that present difficulties, and not needing to bother unduly if machines are idle, as long as any paid operative can be fully and gainfully employed. Often, such a company will have little, if any, rent to pay and the small premises will not attract large rates.

The full impact of high capital costs is felt when machines are acquired on borrowed finance, when full labour costs for all operatives have to be paid out of cash flow, and when premises are large, adequate, and of good standard. If, in such circumstances, the management cannot afford to be selective in their choice of work and have to tackle the difficult, time-consuming jobs, as well as the less critical ones, cost savings of whatever type and magnitude are important.

There are various ways of financing the purchase of injection moulding equipment, but, at a time when borrowing is expensive, an allowance of about 15 per cent per annum for servicing the capital involved would appear to be prudent. In times of financial stringency, such as have been experienced in the United Kingdom since 1970—and to some extent ever since the end of the Second World War—a company is fortunate indeed which has sufficient new capital accruing from profits to finance the purchase of new machines in order to expand a business. It would also seem that, although charges for amortization of equipment are included in all company costings, for some reason or another the capital is never available to replace a worn-out or written-off machine. Frequently, accumulated capital that should have been set aside for buying new machines has to be spent on repairs to keep a worn-out machine operating so that a contract may be fulfilled.

If capital is not directly available for the purchase of equipment, either from accrued profits or from the issue of shares, medium-term finance may be obtained by means of bank loans, mortgages, sale and lease-back of property, hire-purchase contracts, or equipment leasing.

Bank loans for periods over 12 months are now more readily available than in the past, and will be secured by a direct charge upon the company's assets. Such loans, known as 'term loans', are better than taking up overdraft facilities as there is no danger of their being prematurely recalled. Loan charges vary according to the state of the money market but will be 2–5 per cent above the bank's minimum lending rate.

Mortgages may be available from various sources, such as insurance companies, investment companies, or public and private trust funds. Terms may vary from 5 to 20 years, and the money is secured by a charge on the company's assets. Generally, mortgages prove to be expensive. High rates of interest may be charged, and in the event of default the mortgagor may take possession of the asset.

The sale and lease-back of property is a useful method of obtaining additional finance for purchasing additional equipment. Property such as land or buildings is sold to a property investment company under an agreement which allows the property to be leased back to the business for a fixed term with a written-in option for renewal. It has to be borne in mind that the fixed charge for rent will have to be met, irrespective of whether profits are high or low. Property investment companies are not likely to relinquish possession once the sale has been made and this method of financing has that disadvantage. This compares unfavourably with taking out a mortgage, where it is possible to pay back extra capital in good years and so reduce the term of the loan. Sale and lease-back has the advantage that it only reduces the value of the fixed assets by the difference between the value of the property before sale and the value of the lease. This could show capital appreciation in times of inflation or of increasing property values.

Procuring equipment by hire purchase has the advantage of enabling the hirer to use the proceeds of operating the equipment to pay the hire-purchase charges. With injection moulding equipment, for which an amortization period of five years should be allowed, the hirer would be wise to take out his hire purchase over three years so that the written-down value of the equipment will always be greater than the amount of debt outstanding. Hire purchase is expensive because a flat rate of interest is charged for the full amount of debt over the whole of the period, and this virtually doubles the nominal rate of interest which appears to have been charged. The goods purchased under a hire-purchase agreement remain the property of the vendor until the last instalment has been paid, and in most cases maintenance or repairs are the vendor's responsibility. The amount of debt outstanding will appear as a liability on the hiring company's balance sheet.

Credit sales are not subjected to the same legal formalities as hire-purchase agreements and the equipment purchased becomes the property of the purchaser at the time of sale. However, real interest rates are likely to be considerably higher than the nominal rates charged.

Equipment leasing is similar to the leasing of property and is becoming more widely used in the United Kingdom, although it has been in use elsewhere for many years. Equipment is bought by a finance company—sometimes by a bank—and leased for a fixed term to the business. The equipment will always remain the property of the lessor, who will probably base the lease charge on a three-year term. For example, for an injection moulding machine costing £20 000, the annual leasing charge would probably be about £7000. By leasing equipment, a business is able to function with limited capital resources, but the equipment must be capable of earning considerably more than the lease charge. An advantage of the system is that at the end of the lease term the equipment can be replaced by brand new equipment on similar terms and without capital expenditure by the lessee.

Compared with financing by direct loan, equipment leasing is rather more expensive. Thus, for the injection moulding machine leased over three years, the cost would be £21 000. If a bank loan serviced at 15 per cent were used to buy the same machine, allowing 20 per cent per annum for writing down the value of the asset, the cost over a three-year period would be:

	Interest	Amortization	Total
1st year	£3000	£4000	£7000
2nd year	£2400	£3200	£5600
3rd year	£1920	£2560	£4480
		Total	£17 080

and the asset would have a book value of £10 240. Assuming that £21 000 had been allowed for financing the transaction, the difference—£3920—could be subtracted from the book value, leaving a debt of £6320 on a machine having a book value of £10 240. If the machine had been carefully used and regularly serviced it would still be capable—probably for another seven years—of earning at the same rate as when new. However, in an inflationary situation, allowance must be made for the fact that new equipment will cost more than the old equipment did when new.

Various organizations are available for the financing of the purchase of equipment. Among the best known are the Finance Corporation for Industry, the Exports Credit Guarantee Department and the National Research and

Development Corporation. The last is more concerned with financing new developments rather than supplying funds for an established process.

Having dealt at some length with the most important item of fixed cost, that is, the servicing and obtaining of capital, consideration should be given to the other cost items in running an injection moulding business. It is convenient to categorize these into three groups:

(1) Costs that tend to be constant, irrespective of whether output is high, low, or non-existent.
(2) Costs that vary directly with output.
(3) Costs that occur on a variable basis, but are not directly related to output.

Among the items in the first group to be considered, other than the servicing of capital, are:

(1) Rent of property, rates, water rate (if fixed), and allowances for repairs, painting, decorating, etc.
(2) Heating, cleaning, and lighting, and the maintenance of services.
(3) Maintenance of stores of raw materials, components, moulds, and spare parts, and for bought-in items, such as inserts.
(4) Fixed allowances for the repair of machines, moulds, and services on a planned maintenance basis.
(5) Telephones, Telex services, timekeeping equipment, fire alarm and prevention services.
(6) Office overheads; allowances for stationery, typewriters, photocopiers and other office equipment.
(7) Indirect labour, such as management, accountants, secretaries, draughtsmen, storekeepers, etc., and toolroom personnel, if employed.
(8) Allowances for sundries, such as canteen, restroom, first-aid kits, beverage vending equipment, etc.

Of the fixed costs listed (which may be by no means exhaustive) item (1), together with the servicing charges for capital, will be dealt with separately in working out costings. The other items will normally be lumped together under the general umbrella of overheads. However, anomalies may arise if this is done, as will be seen below when discussing the true cost of material.

The main variable costs, that is, those costs that vary directly with output, are the cost of raw material, allowances for mould amortization, and the cost of packaging material for the dispatch of finished components. In determining these costs, it is important to make some adjustments unless a costing system is used which already does so. These include:

(1) Raw material cost is *not* the price of the material as paid to the supplier, but the supplier's price plus the cost of storage, the cost of labour and machinery used in storing the material, the loss of interest on capital, and the cost of servicing the capital used in paying for the material. Thus, if a supplier

allows six weeks credit in settling the account, but a three months stock level must be maintained, the cost of financing for six/seven weeks and the cost of storage for three months must be added to the basic price. If a labourer is used for moving material in and out of stores, allowances must be made for his wages and overheads. Allowance must also be made for the effects of inflation on the cost of raw materials.

(2) Mould amortization can be justifiably considered as a variable cost, because the mould cost per item depends on the length of run and the time over which amortization is to be made. Some moulds need more servicing than others (for example, fast cycling moulds) and estimated costs for servicing must be included under this heading.

(3) The cost of packaging material for the dispatch of finished mouldings also varies with output and should be dealt with as a variable cost item. Some injection moulding companies have run into financial difficulties over contracts because they have failed to add realistic amounts for this item. Adding a fixed percentage to a quoted price to cover the cost of despatch may result in either too high or too low a quotation.

Some of the semivariable costs, that is, those costs that vary from time to time but in a manner not directly related to output, include:

(1) Cost of electrical power. A certain amount of electricity is used in the factory irrespective of whether or not the full productive capacity is being maintained.
(2) Cost of some labour such as casual labour or outworkers recruited during periods of high output.
(3) Sales costs such as sales persons' salaries and commissions, travelling allowances, and the cost of samples and advertising.

Costs of this type may be difficult to determine accurately in respect of any given job and are generally treated as an item of overheads. Provision should, however, be made for including this type of overhead into job estimating on a variable percentage basis to avoid either over- or under-charging.

One further item of variable or semivariable costing which should be mentioned is the cost of reclaiming scrap material. As with some other items, it is all too easy to overlook this item in the belief that any material reclaimed effects a saving of the full price of the raw material. This is not so. Even if an operator has a small granulator by the side of the machine, used for grinding sprues and runners as they are made, there is still a significant amount of labour involved. In addition, there is the capital cost of the granulator to be serviced, the cost of its maintenance, and the cost of the electricity consumed to be added—not to mention the cost of additional cleaning necessary. At best, these may add up to one-tenth of a man-week per tonne of material reclaimed, but if a man is employed full time on reclamation work, depending on the size of the granulator, the cost of the reclaimed material may be from $\frac{1}{4}$ to $\frac{1}{2}$ man-week in terms of wages (at 1978 prices, £20–£80 per tonne).

Budgeted machine hourly rates

A convenient method of job-costing in injection moulding involves the working out of a machine hourly rate for each machine or group of machines in the moulding shop. In this respect it is best to choose machines of similar age and capacity in deciding on the groups. It may then become possible to quote keener prices for less critical jobs. In most moulding shops there are machines which are 'past their prime' but which are still capable of producing reasonable quality mouldings. Although general overheads will be at the same level as for newer machines, the fact that their value is written down to a low figure means that lower allowances need be made for the servicing of capital and a lower machine hourly rate (MHR) could be charged.

In working out the budgeted machine hourly rate, the fixed costs are first computed on a yearly, monthly, or weekly basis. The planned production week is then decided upon—single-shift, two-shift or continuous working—and the total number of working hours is fixed. Total fixed costs are then divided among the machines in terms of the capital value of each machine, for the period decided upon, and the fixed cost for each machine is divided by the total number of hours in the period. This period may be 50, 80, 100, 120, or 150 h per week or 2500, 4000, 5000, 6000 or 7500 h per year. Few moulding shops work completely continuously, and it is often found best to close down at weekends and use the period of shutdown for essential maintenance. A year of 50 weeks is also used because most factories close for a week or more at Christmas and may have another week during the year when maintenance can be done.

A realistic estimate is then made of the total possible productive time as a percentage of the total time the factory is open. With single-shift working this could be 80 per cent since, in a 10 h day, 5-day week, it is only too easy to lose 2 h per day on each machine. Thus the machine hourly rate becomes:

$$\text{MHR} = \frac{\text{Total fixed costs} \times 100}{\text{Total hours worked} \times 80}$$

Even with continuous working over, say, a 120 h week, it would be safest to budget on about 100 productive hours, because problems can occur at times of shift changeover and minor breakdowns can reduce efficiency by another 10 per cent. There may also be mould changes for which no charge can be made on the customer, and although these should be minimized, allowance must be made for their possible supervention so that the MHR is not fixed at too low a level.

Having decided upon the total number of working hours for a given period and estimated the MHR, a period of trial is necessary during which a careful check is kept on the output of the machines and the total productive time. If the budgeted productive time turns out to have been too generous, the MHR will need to be adjusted or, perhaps better, the reasons for the lower-than-budgeted productive time investigated. It is all too easy to adjust a costing figure when what is really needed is a purge on inefficiency.

In budgeting machine hourly rates, some cost accountants include all fixed and semivariable costs, using a figure for overheads that would include such items as electrical power, labour, mould amortization, etc., in fact, all costs except that of the moulding material. This can give rise to distorted figures for machine hourly rates making realistic job-costing difficult. In fact, the large variations often found in the industry when cross-checking from moulder to moulder on a component price are probably due entirely to different methods of estimating machine hourly rates.

Estimating and pricing of component production

Having arrived at a realistic figure for the machine hourly rate, the estimator has to decide:

(1) Which machine will be used?
(2) How long will it take to mould the component?
(3) What is the true cost of the raw material?
(4) What percentage should be added on for profit?

The first of these is determined by the size and weight of the component, the number of impressions, and the size and complexity of the mould. There may be some overriding reason for using a particular machine, such as a very deeply drawn article requiring a machine with a very large daylight, or a much larger machine than its shot weight would seem to justify. All such matters must be considered when deciding on the machine to use but, perhaps of even greater importance with the development of some of the newer thermoplastics, is the question of whether the machine is capable of handling the material involved.

The second question is largely a matter of arithmetic and a knowledge of how rapidly the machine will be able to operate, bearing in mind the characteristics of the mould. Having worked out the cooling time for the material (generally the major part of the time cycle), using the formula discussed in Chapter 4, the machine 'dry cycle' time is added, adjustments having been made for making sure that mouldings have fallen clear of the mould and that undue wear and tear will not occur because of too rapid opening and closing of the mould. The component cost, as a proportion of machine hourly rate, is then:

$$\text{Component cost} = \frac{\text{MHR} \times \text{cycle time (seconds)}}{3600 \times \text{number of impressions}}$$

Determining the true cost of the raw material is slightly more complex. Some moulders include the cost of storage, handling, wastage, and reclamation of material in their factory overheads and include a proportion of these costs in their machine hourly rates. This course of action can lead to some distortions in component costing between components using large quantities of material on a short moulding cycle and those using small quantities on a long cycle. Furthermore, the cost of storage and handling of an expensive material expressed as a percentage of price is much lower than it is for a cheaper plastic.

Consider two jobs, one using 10 tonnes of material per week at a price of £400 per tonne, the other using 1 tonne per week at £1400 per tonne. The main items of overheads are interest on capital, cost of storage, and handling. Assume that the stock level to be maintained is four week's supply of material and that capital service charges amount to 15 per cent per annum. Storage costs would probably be about £1 per square foot and labour charges about £80 per week.

Job No. 1: 10 tonnes/week, £400/tonne

Capital outlay £16 000. Interest for four weeks @ 15%	£185.00
Storage 400 ft^2, 10 ft high	£ 46.00
Labour: $\frac{1}{4}$ man for 4 weeks	£ 80.00
	£311.00

Cost per tonne £311 ÷ 40 = £7.78

True price of raw material £407.78/tonne, an increase of 1·95 per cent.

Job No. 2: 1 tonne per week at £1400 per tonne

Capital outlay £5600. Interest for four weeks @ 15%	£64.60
Storage: 40 ft^2 at 10 ft high	£ 4.60
Labour: $\frac{1}{40}$ man for 4 weeks	£ 8.00
	£77.20

Cost per tonne £77.20 ÷ 4 = £19.3

True price of raw material £1419.3/tonne, an increase of 1·38 per cent.

In deciding how much to add on for profit, due regard must be made to the particular job and the customer involved. Profit margins can be reduced if the job will bring prestige to the company, if it will help to gain another and more lucrative contract, or if there is spare capacity. However, it would be unwise to decide such matters on an *ad hoc* basis. In preparing a yearly budget, it is prudent to allow a certain sum for promotional activities, quite apart from a normal publicity budget, and in deciding on the pricing of any contracts which have a promotional bias, to allocate a proportion of this sum.

Cost saving methods

So far in this chapter the emphasis has been on paying careful attention to the financing and costing of projects, and it is in this field that the greatest contribution to profitability can be obtained. There are, however, some other methods by which cost savings may be achieved. These include rationalization of material stocks and grades, utilization of standard material, and the use of master batches, etc.

Carrying large stocks of a large variety of different materials in a wide range of grades and colours can be extremely expensive and is to be avoided if possible. If it is necessary to obtain supplies of a special grade or colour for a particular job, careful estimating of the quantity required will show savings in several ways. First, in having the minimum surplus material to dispose of at the end of the contract. Second, in direct saving of interest and storage costs, which, on a medium-priced material could be as much as 12 per cent per annum on the manufacturer's price (of course, in a highly inflationary situation this can show a profit). Third, because stores do not expand and contract as material comes in or goes out, the carrying of unnecessary stocks causes congestion, and so wastage of the storekeeper's time.

Any obsolete material should either be used or disposed of as soon as possible. There would seem to be little point in keeping sizeable quantities of, say, nylon at £1400 per tonne in the hope that one day a use will be found for it. The disposal price can generally be discounted to allow for the interest and storage costs which would be incurred if it were kept in stock.

Similarly with finished components: if a contract ends with unsold components, steps should be taken to dispose of them. Keeping unpaid-for components in the goods outward department of the store awaiting the customer's call-off is extremely foolish. Not only are there interest charges on the raw material to consider, but there are also charges on the work that has gone into the moulding of the parts. Taking the previously stated figure of 12 per cent per annum for raw material storage costs, for finished components this could easily be 35 per cent per annum. Moulded articles generally take up a vastly greater space than raw materials, and it is all too easy for the moulder who has adequate storage space to allow customers to leave their finished components in the store 'until they can be collected', and then to find that all the raw material storage space has been used up as well.

The same principle applies to scrap mouldings to be re-granulated. On occasion these have been allowed to accumulate, possibly because of a breakdown of the granulator or a shortage of labour to operate it, to such an extent that it becomes difficult to find space for new material or for good quality finished articles. It must always be remembered that storage space costs money, which must be used to the best advantage, and that any material standing in the stores is using up capital resources. It is a very sound principle to work so that all scrap material fit for re-work should be granulated, dried, packaged, and labelled ready for use as soon as possible. Storage of such scrap, often in inadequate packages, allows it to accumulate dirt and moisture which is then difficult to remove.

If a standard material is to be used in a number of colours, the master batch techniques should be employed. Many materials are now available from the manufacturers in the form of natural coloured powders, with little except stabilizers by way of additives. They demand a slightly lower price than compounded materials, that is, than materials with colour or other additives incorporated by a process involving fusion and granulation. Because they have not undergone the compounding heat treatment they can be considered as slightly better-quality materials, in that colour (especially of acrylic plastics) may be slightly better and flow properties, such as melt flow index, may be more uniform.

These materials are coloured by adding a colour concentrate, known as a masterbatch, which can either be bought from the manufacturers or from one of the many specialist companies now in this sort of business, or they can be made up by the user. Pigments suitable for making master batches are now readily available from dye and pigment manufacturers. Many material manufacturers publish or supply formulae so that their customers can make up and match a standard range of colours, and some provide a colour matching service. The quantity of colour concentrate added will be between $\frac{1}{2}$ and 5 per cent, depending on the type of colouring matter and the density of the colour required.

The coloured master batch is intimately mixed in the correct proportions with the natural powder in a tumble blender, which consists of a drum or container rotated end-over-end. Its volume should be about $1\frac{1}{2}$ times that of the material, so that a good free space is left for the tumbling to take place effectively. Baffles to assist in the mixing are attached to the walls of the container. After about 15–20 min mixing, the blended powder is fed to the injection moulding machine, where the rotation of the screw during fusion helps to produce a homogeneously coloured melt.

The method adopted when using pigments to match the colour is to make up a quantity of 'super concentrate' by mixing the pigments with the powdered plastic. This is adjusted to give the correct colour match on dilution and from the concentrate supplies of masterbatch can be produced. It is found easier to weigh out or to measure the exact amount of colour when it is mixed with plastic than when the pigment is in its concentrated form. If a considerable quantity of the super-concentrate is made up, it is possible to make supplies of masterbatch with little batch-to-batch variation.

By using this technique, savings of several pounds sterling per tonne of material can be achieved and the apparatus required is simple and inexpensive. Colour mixing to a not-too-rigid specification working to a supplier's formula is quite well within the capabilities of a normal operator, but for more stringent colour specifications it is best to delegate the work to someone who has been given training in colour matching.

In addition to using coloured master batches it is possible to apply the technique to the adding of fire retardents, special stabilizers, or mineral fillers, such as talc, kieselguhr, and some types of glass filling.

A variation of the master batch technique known as 'dry blending' employs natural granules instead of powder. The pigment, either in concentrated form or mixed with powdered polymer, is tumbled with the granular material and adheres to the surface of the granules. On being mixed and fused in the injection cylinder it gives a homogeneously coloured melt. A homogeneously coloured melt can also be obtained when using granular master batches and natural granular material in a screw machine. This is, in fact, how the master batch technique was first started. However, there is always the possibility when granular materials are used, of obtaining slight streakiness of colour and the method would not be recommended, for instance, for the incorporation of carbon black into polyolefines for out-door weathering grades of black.

If the granular master batch technique is used for moulding on plunger machines, a very streaky appearance of the moulding will result. However, some moulders, wishing to simulate the appearance in the mouldings of mahogany or onyx, use blends of different coloured granules in plunger-type machines and obtain very pleasing patterns. Like the natural counterparts, no two pieces have exactly the same pattern.

The cost savings already mentioned come about in several ways. Firstly, when buying large quantities of a single type or grade of material, quantity rebates operate. Secondly, it is often cheaper to employ labour and the inexpensive machines needed in a small organization than in a large company carrying higher overheads. Thirdly, the exact quantity of coloured material required can be made up with no surplus to give waste. Finally, because stocks of different colours are not being held in the store, space lost by the need for segregation is reduced to a minimum.

Optimizing time cycle and machine productive time

Whatever ideas for cost saving are introduced into the injection moulding of plastic components, ultimately the time cycle of the machine and the rate at which it can be made to produce good saleable mouldings are the criteria for profitability. Unfortunately, the operator or setter who does not fully understand the working of the machine will try to snatch the odd second here and there in the time cycle, often reaching a stage where either faulty mouldings are being produced or where, in multi-impression work, one or two impressions are not operating properly. The continued production of faulty mouldings on, say, one impression of an eight-impression mould can soon reduce profitability to a very low level or even cause the machine to be running at a loss. It is necessary to consider not only the loss of one-eighth of the production, but also the cost of the material being used, the cost of segregating the bad mouldings, and the cost of re-granulating the scrap material. There is also to be considered the possible wear on the mould if, for example, flashing is occurring, or of causing mould distortion if one cavity is not being filled. This latter practice can put the mould locking out of balance.

There is a strong case for setting a very rigid machine time cycle and some moulding companies do this by having the timers locked after they have been set. This may have disadvantages because, with the slightest variation there is a need for the operator to call the setter (who may be urgently employed elsewhere) to make the adjustment. The use of electronic sensing and feedback equipment may ultimately minimize the need for manual adjustment, but there will probably always be the need for properly trained operators or setters to carry out the 'fine tuning' of a machine to obtain optimum output.

In the analysis of the injection moulding cycle in Chapter 2, the main elements of the process and the time taken for each element were discussed. The minimum time for each element is fixed either by the capabilities of the machine or by the properties of the material being moulded. There are, however, machine functions falling between these main elements which can be a cause of loss of time.

Consider, first, the locking of the mould ready for injection. This must be carried out swiftly but gently. The speed is controlled by a flow control valve in the hydraulic fluid line and the mould protection by a microswitch which only allows the high pressure for locking to be applied when the mould parts are virtually in contact. If the settings are not made accurately, the mould may take longer to close than necessary and valuable time is lost.

The next phase is injection. The minimum time for injection is fixed by the rate at which the hydraulic pump will supply fluid or by the rate dictated by the nature of the material being moulded and the configuration of the mould. But injection cannot take place until the cylinder nozzle is in contact with the mould sprue bush. If the machine is designed so that the closure of the mould gives the signal for the cylinder carriage to move forward, time can be lost if the carriage moves slowly and over greater distance than is necessary. (Some machines have very slow-acting hydraulic cylinders to move the carriage, the object being to obviate damage to the mould sprue bush and nozzle.) Slight time delays occur at each stage of machine operation, but the good setter will anticipate such delays, caused by the inertia of the solenoid valves in the hydraulic system, and will carry out the fine setting accordingly.

If the machine is operating with sprue break, that is, the withdrawal of the nozzle from the mould sprue bush after each shot, the minimum of carriage movement should be allowed, so as to reduce the loss of time before injection can take place. The actual time during which sprue *break* occurs overlaps either cooling or the start of mould opening.

Unless a long plunger dwell time is required, as, for example, in the flow moulding of thick sections in some materials or the moulding of thick sections in nylons and acrylics, screw-back or plasticizing should start immediately the screw/plunger reaches the end of its forward stroke. Especially with high-speed injection through a properly designed gate, there is little point in maintaining a long plunger dwell because the gate freezes immediately injection stops. The screw-back time should not be longer than the additional cooling time required and will rarely be so in normal moulding practice. At the end of screw-back or cooling, whichever is the later, mould opening will be signalled.

The same criteria apply to mould opening as to mould closing, the mass of metal being controlled in its movement to minimize damage to the machine.

During mould opening, ejection of the moulded parts takes place. When using a mechanical system, as the mould opens the ejector bar comes up against a stop or pin fixed to the machine body and the ejector mechanism is actuated. Too rapid a platen movement at this stage can cause bad jolting of the machine, undue wear on the ejector system, and the ultimate breakdown of the machine or mould. Fine setting of the actuating gear is necessary to achieve speed with minimum strain. One advantage of hydraulic or pneumatic ejection systems is that they can be set to operate with very little jolting and can be timed so that ejector movement is complete and the moulding ready to drop from the mould at the instant that the mould becomes fully opened.

Before the commencement of the mould closing operation it is necessary to allow time for the mouldings to fall clear of the mould. On all machines operating fully automatically a time delay switch is fitted to give a suitable 'pause' before the signal to re-start the cycle is given. Frequently, the pause timer or interval timer is set to allow more time than is absolutely necessary. If it is assumed that the moulding will fall free of the mould by gravity (ignoring air resistance, which will be very low except in the case of large-area lightweight mouldings), the pause time required can be calculated approximately as follows:

$$t_p = \frac{\sqrt{d}}{22}$$

where t_p is the pause time in seconds
d is the distance in centimetres from the topmost point of the moulding as it fits in the mould to the lowest part of the mating surfaces of the mould or of any other obstruction on which it can be caught as the mould closes.

Thus, for a moulding that has to fall 30 cm to be clear of the mould the pause time will be

$$\frac{5{\cdot}5}{22} = \tfrac{1}{4}\,\text{s}$$

A more detailed chart of the cycle of operations is shown in Fig. 12.1. 'Productive time' indicates the period during which material is actually being handled in the mould and 'nonproductive time' is the additional time for the machine to complete its full cycle. Obviously it is desirable to keep the ratio $AB{:}(CD + EF)$ as great as possible, bearing in mind that even during the nonproductive time material is being heated in the cylinder. However, the minimum time AB is determined by the nature of the material being moulded, so having settled the optimum conditions to produce desirable quality components (and hence the minimum productive time), the setter should then reduce the nonproductive time to a minimum. The limit on this will be the wear-and-tear on the machine, which should always be a consideration.

Slight time savings can be effected by anticipating some of the signals but

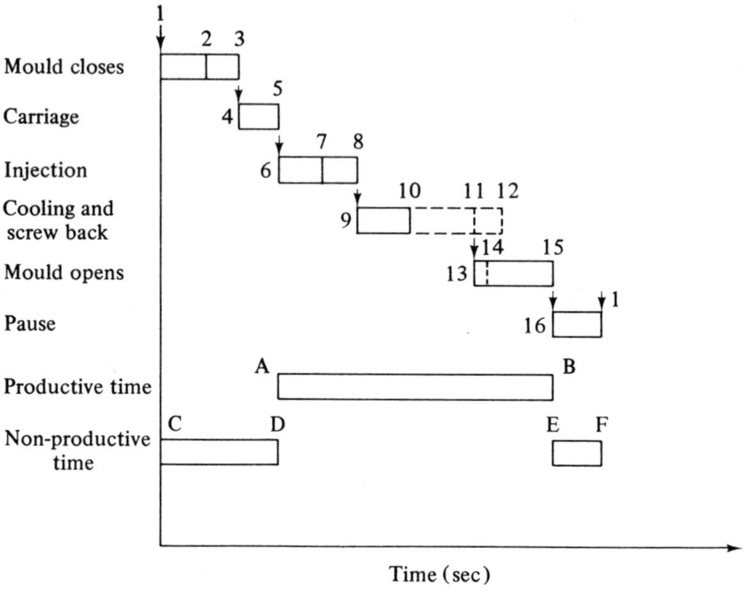

Fig. 12.1 Operation cycle chart

this must be done with great care. Solenoid valves energized through microswitches or timing devices do not respond instantaneously. There is a very small but significant time lag. Thus, by observation, the setter using a pause timer set for 0·5 s could reduce this to, say, 0·25 s and by carefully watching the mould locking could see how another 0·25 s could be saved. By studying the different elements of the cycle and making the necessary fine adjustments, the setter can make a significant contribution to increased profitability. One second clipped off a 30 s cycle on a single-impression mould on a machine with an MHR of £6 can show a saving over a 24 h working day of nearly £5. The same time taken off a 5 s cycle with the same machine hourly rate saves £36 per day!

Table 12.1 can be used for calculating the hourly production rates for single impressions with cycle times of 1·0–9·9 s (or 10–99 s by moving the decimal

Table 12.1 Hourly production rates at cycles of 1–10 s

	0·0	0·1	0·2	0·3	0·4	0·5	0·6	0·7	0·8	0·9
1	3600	3272	3000	2769	2571	2400	2250	2117	2000	1894
2	1800	1714	1636	1665	1500	1440	1384	1333	1285	1241
3	1200	1161	1125	1090	1058	1028	1000	972	947	923
4	900	878	857	837	818	800	782	765	750	734
5	720	705	692	679	666	654	642	631	620	610
6	600	590	580	571	562	553	545	537	529	521
7	514	507	500	493	486	480	473	467	461	455
8	450	444	439	433	428	423	418	413	409	404
9	400	395	391	387	382	378	375	371	367	363

point one place to the left), and illustrates the significance, particularly with short cycles, of a difference of 0·1 s. Even at 10 s the difference in hourly rate is four mouldings (at an MHR of £6 this is 6·6 p/h), whereas at 2 s the difference is 90 mouldings, equivalent to 30 p/h.

Effects of process variables on profitability

To compare the effects of process variables on profitability it will be necessary to make certain assumptions and to formulate a hypothetical case study. Although actual figures may not be very meaningful, such a study will indicate the trends in profitability as efficiency increases, reject rates decrease, time cycles are tightened, and material prices brought to a minimum.

The first assumption made in the case study is that capital (£170 000) is borrowed at an interest rate of 10 per cent. This is probably too low a rate during periods of inflation, but may be fairly realistic when taxation is taken into account. However, for the purpose of this case study, taxation and capital allowances are ignored. It is also assumed that provision will be made for the repayment of the capital in five years. The figures in the case study will be for the first full year's trading and would need to be adjusted for the second and subsequent years.

A second main assumption is that the selling price of the articles will be twice the cost of the raw material. This is a rule-of-thumb method of working out an article price often used in quick mental calculations. It applies to small articles in the range up to about 200–300 g.

The third assumption is that profitability is shown most meaningfully as a percentage gross annual return on investment. This does not take into account such items as general company overheads, profits, and dividends, which will, of course, vary from company to company.

Having borrowed £170 000 at 10 per cent per annum, the company in the case study buys five injection moulding machines in the medium price range at, say, £20 000 each. Ancillary equipment and provision for services cost another £35 000, leaving £35 000 as liquidity for paying wages, buying material, etc. Premises will be rented and will comprise 3000 ft^2, including a small office, toilet accommodation, etc., at £1 per ft^2. An area of 500 ft^2 will be used for a raw material store, and the same area for finished products awaiting despatch. Five weeks stock of material (50 tonnes) will be maintained and the finished products store will accommodate one week's production (at five times the raw material volume). The machines will take up about 1000 ft^2, leaving 1000 ft^2 for offices, etc., mould storage, and a small work bench for mould repairs.

Material would normally be bought at a price which, after allowing for storage and handling costs, would come to £400/tonne and it is assumed that all the machines produce the same articles, for which 8 p each could be charged. The minimum of labour would be employed but, for three-shift working of 120 h per week, it is necessary to have seven persons who can all work at all hours. There would be two per shift and one spare person, say, 4–5 operators and 2–3 fitter/setters. Two supervisory staff would be the minimum necessary, doubling as production manager/sales/delivery/progress chaser, etc.

Power requirements for the small machines would be about 100 kW at an approximate price of 1·5 p per unit.

Fixed costs include:

Interest on capital (first year)	£17 000
Rent on 3000 ft^2	£ 3000
Rates (estimated at $\frac{2}{3}$ rent)	£ 2000
Provision for repayment of capital	£34 000
Labour and supervision	£35 000
	£91 000

Among variable costs are:

Material at £400/tonne
Power at 1·5 p/unit

Semivariable costs have not been included. These are items such as administration, accountancy, sales, delivery, telephone services, cleaning, etc., but they will not affect the trends shown in the study, only the detail.

The selling price of the articles will be 8p each however many are sold, and the standard moulding cycle 20 s. Component weight is 100 g.

Effect of machine utilization. Figure 12.2 shows the effect on percentage gross realization on capital of machine utilization. It assumes that 100 per cent satisfactory mouldings are being made during the whole of the operating time. The percentage machine utilization is:

$$\frac{\text{Number of productive hours} \times 100}{\text{120 (number of budgeted hours)}}$$

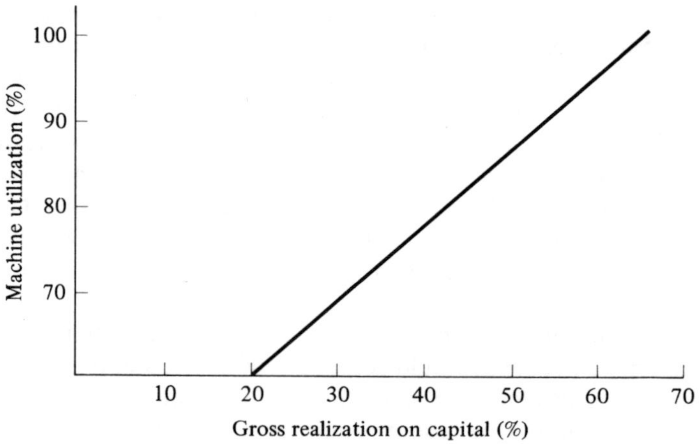

Fig. 12.2 Percentage machine utilization/profit

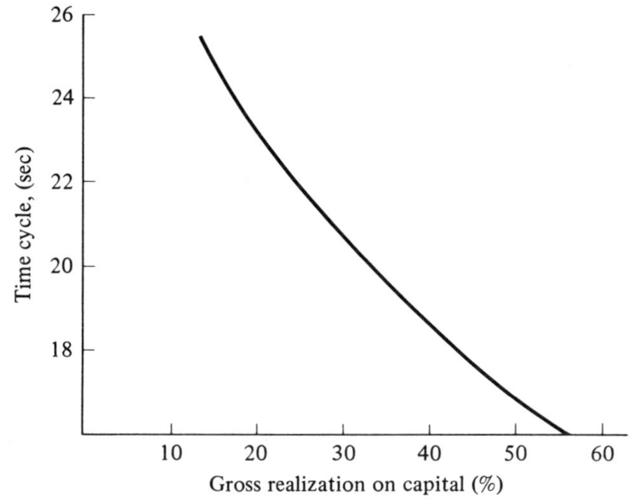

Fig. 12.3 Time cycle/profit

Effect of cycle time. Figure 2.3 shows the effect on percentage gross realization on capital of changes in the cycle time at 75 per cent machine utilization. It assumes 100 per cent satisfactory mouldings.

Effect of material price. Figure 12.4 shows the effect on percentage gross realization on capital of changes in the cycle time, again at 75 per cent machine utilization. It will be seen that in this case study a change of £20/tonne approximates to a change in cycle time of 2 s (10 per cent).

Cost of producing faulty mouldings. Figure 12.5 illustrates the effect on percentage gross realization on capital of producing faulty mouldings at 75 and 100 per cent machine utilization. The value of the scrap material and the

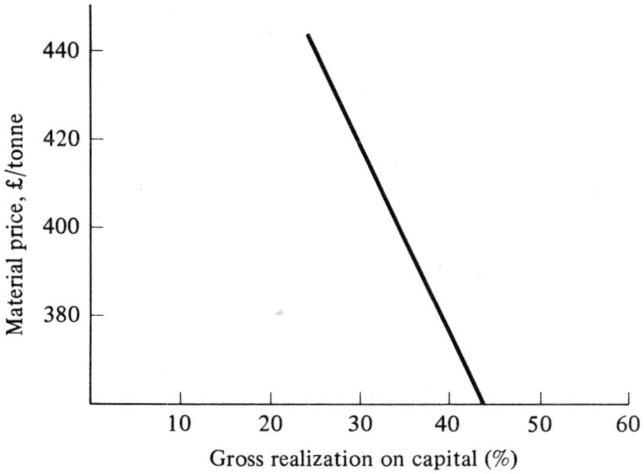

Fig. 12.4 Material price/profit

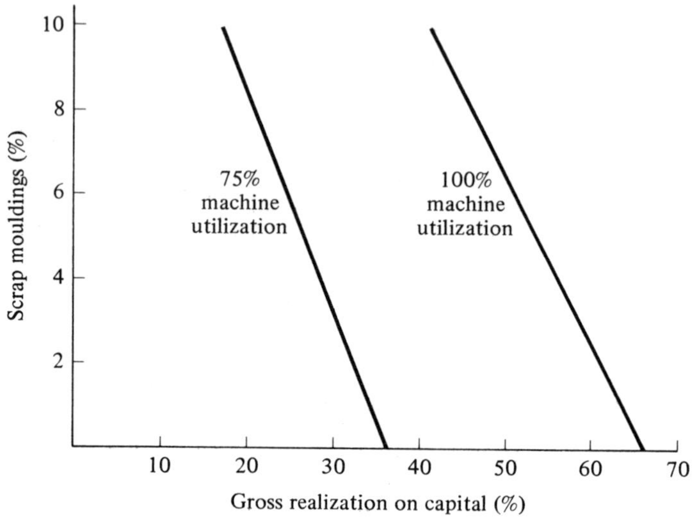

Fig. 12.5 Percentage scrap mouldings/profit

cost of reclaiming are not included in the calculations. It will be seen that the higher the percentage machine utilization, the greater is the cost of producing scrap. With low machine utilization and an already inefficient organization, the production of scrap mouldings would not appear to be so important. However, it should be observed that at 75 per cent machine utilization and 18·7 per cent scrap production, the percentage gross realization on capital falls to zero.

Conclusion

There has been a good deal of emphasis on efficiency in this book and in particular in the final chapter, in which some of the financial problems of the injection moulding industry have been discussed. It would appear that financial success in this highly technological branch of production engineering arises from three factors: (1) Thorough understanding at all levels of management and production of the principles involved in the injection moulding process; (2) attention to detail in the operation of machines and the general conduct of the moulding shop; and (3) careful financial control in the buying of raw material, the selling of finished products, and the utilization to the maximum of the highly expensive capital equipment used.

At all levels in the injection moulding industry there is a shortage of skilled, fully trained personnel. There is also often a lack of liaison between component designers, mould makers, moulders and users of moulded components. The first of these can be remedied in time by making use of the facilities offered by research associations, machinery and materials manufacturers, and the Industry's Training Board, and by the encouragement of young people to enter the industry and to undergo courses of study at the many educational establishments offering suitable tuition.

The latter is a matter of attitude. The injection moulding of plastics components is a challenge to the ingenuity, resourcefulness, and intelligence of all those who take part in it. If this volume has been able to present some of the problems in a logical manner and to indicate some possible ways in which they can be solved, its purpose will have been achieved.

INDEX

Accumulator, 2, 39
  in mould filling, 112
Acetals, 69
  signs of degradation, 132
After-shear swelling, 85, 86
Alkyd plastics, 122
Aminoplasts, moulding, 122 ff.
Amorphous plastics, 71
Appearance surface, 2
  specification criterion, 14
  faults in, 135
Atactic, 68

Back-flow valve (*see* Check ring) 2
Backing plate, 2
Bakelite, 19
Balance tray, 2
Barrel (*see* Cylinder) 3
Blistering, of thermosetting mouldings, 128
Bolster, 3
British Plastics Federation, 13
British Standard 1755, 2
Burn mark, 9
  causes and cures, 130–132

Carriage hydraulic cylinder, 51
Carrot, 3
  in 3-plate mould, 103
Cartridge heater, 3
  in hot runner moulds, 104
Cavity, 3
  in mould, 91
  preferential cooling, 106
Cavity register, 3
Check ring (*see* Non-return valve) 3, 29, 47
  effect on mould filling, 115
Co-axial screw machine, 26
Colour changing, 152
Colour-coding of pipes etc., 158
Colour, effect on heating of material, 80
Component design, 106
Compound, 3
Compressibility of plastics, 81
Compression ratio, 3
  for screws, 47
Computer control, 36
Conductivity, thermal, of plastics, 75, 76

Control:
  of mould temperature, 176
  of oil pressure, 43
  of time, temperature etc., 61
Control systems, 176
Coolant fluid, precautions, 151
Cooling water supply and requirements, 172–174
Copolymer, 3, 70
Core and core pin, 3
Core, preferential cooling effects, 106
Costs, 182 ff.
Cracking or crazing, 9
  cause and cure, 132
  splay marks as cause, 141
Crystalline plastics, 71
Cuff heater, 3, 46
Cycle of operations, 21, 30
Cycle time, 3
  calculating, 80
  optimizing, 193
Cylinder, 3
Cylinder temperatures:
  for thermoplastics, 79
  for thermosetting materials, 125

Daylight, 3
Decomposition, as cause of splash marks, 140
Decomposition temperatures, 77
Decompression (suck-back in nozzle), 31
  as cause of burn marks, 131
Decompression zone, 3
De-gating, 3, 97
Delamination, 10
  cause and cure, 132
Density of plastics, 72, 79
Depolymerisation, 3
Dice cut, 3
Differential shrinkage, 105, 143
Distortion, cause and cure, 10, 133
Distortion temperatures, 79
Dowel pin, binding on bush, 56, 88, 120, 133
  (also guide pin), 3
Draft and draft angle, 4
  as cause of scuffing, 136
Dry-blending, in cost saving, 192
Drying equipment and temperatures, 179

Ejection of mouldings, 21, 94
Ejector pin and plate, 4

Ejector system, 4, 58
Energy regulator, 62
Enthalpy of plastics, 74, 75
Extended nozzle, 4, 104

Family moulds, 102
Faults, causes and cures, 9–18, 130 ff.
in thermosetting mouldings, 128
Fault-finding charts, 16, 147, 148
Feed hopper, 49
Feed setting, 24, 31, 120
Fillers, 125, 136, 192
Financial aspects, 182 ff.
Finished goods record, 161
Fire precautions (hydraulic oil), 40, 43
First aid, 156
Flash, 10, 133
Flash mould, 4
horizontal, 90
vertical, 91
Flow control valve:
in injection line, 51
variations in, 117
Flow mark, cause and cure, 134
Flow moulding, 4
Flow path ratio, 110
Freeze time, 21
Frozen-in stress, 67, 82, 116
mouldings free from, 132
in warped mouldings, 143

Gas marks, in thermosetting mouldings, 128
Gate calculations, 99
Gate, 4
for thermosetting moulding, 127
land length of, 99, 141
shear rate in, 17, 87
sizes v shear rates (graphs), 100
types for injection moulds, 97
Granulator, 177
hazards, 150, 152
Guards for machines etc., 154

Heater band, 2
Heating cylinder, 45, 46
Hopper driers and loaders, 178
Housekeeping, 13, 149 ff.
Hyatt Brothers, 19
Hydraulic ejection, 58, 95
Hydraulic fluid:
contamination, 39
control, 42
emulsions with air and water, 41
nature and properties, 40
pipework, 43
reservoir, 38
temperature, 39, 117
Hydraulic injection cylinder, 50
Hydraulic pressure, 20, 112
Hydraulic pump, capacity and volume, 50, 115

Impression, 4
Injection, 20
power and pressure requirements, 112 ff.
variation, effect on mould filling, 117
Injection pressure, 4
for thermosetting plastics, 126
Injection rate, 108, 109
Insert, 5
moulded-in, 107
Intensification ratio, 5, 50, 51
Isotactic, 68

Jack ram, 55, 56
Jetting, 144
Jig cooling, 5, 133
Job record, 162
Job report, 163

Land, 5
length of in gate, 99, 141
Lift, 5
Limit switch, 5
Line pressure, 5, 50
Locking force, 5
(mould lock), 20
variations in, 119, 120
Locking mechanism, 5, 53 ff.
Lubrication, 151

Machine control, 61
Machine rates, 165, 188
Machine utilization and profitability, 166, 198
Magnetic screen, 49
Masterbatch, 5, 192
Material balance, 163
Material cost (true), 186, 190
Materials for moulding, 65 ff.
cost of reworking, 187
price v. profitability, 199
Mica mark (*see* also Splash mark), 10
Mica marks, causes and cures, 139, 140
Mixing equipment, 177
Moisture, as cause of splash marks, 140
Moisture, maximum permissible for moulding, 179
Molecular structure of plastics, 66 ff.
Monomer, 5, 70
Moulding cycle (*see* Cycle of operations), 6
Mould clamping, 91
Mould cooling, 93, 105
Mould filling, 92
assessing by making short mouldings, 138
power requirements, 110 ff.
time, 109, 110
Mould gates, 96, 98
Mould height adjustment, 6, 53, 55, 57

Mould locking:
effects on mouldings, 117
hydraulic, 55
„ and toggle, 57
mechanisms, 53
toggle, 54
Mould manufacture, 95
Mould opening and closing, 21
Mould platens, 52
Mould release agents, 133
Mould, split, 20
Mould staining in thermosetting moulding, 129
Mould temperature control:
effect on warping, 143
weld lines, 145
Mould temperature controllers, 176
Mould, three-plate and hot runner, 102, 104
Mould unit, 52 ff.
Mould venting, 131
Mould, vertical flash, 91
Moulds:
materials of construction, 88
prototype, 89
for thermosetting materials, 127

Non-return valve:
ball type, 30
check-ring type, 29
on screw, 47
Nozzle, 6, 46
extended, 104
leakage at, 31
Nylon, 69

Ovens, drying, 180

PVC, 69, 132
Pause time, 6, 195
Phenoplasts, moulding, 122 ff.
Pipework, 44, 154
Plant layout, 169 ff.
Plasticizing capacity, 6, 60, 72
Plasticizing cylinder or unit, 6
description, 45
function, 20
Platens, mould, 6, 52
Plunger, 6
Plunger machines, 21–25
Polymer, 6
Polymerization, 6, 70
Polypropylene, 68
Polystyrene, 69
Polythene (polyethylene), 66
Positive mould, 4, 91
Power:
electrical requirements, 168
requirements for mould filling, 112
Power unit, 20, 37
Preheating, 7
Pressure:
control, 43
effect of on plastics, 81
excessive, as cause of sticking, 142
requirements for mould filling, 112 ff.
Pressure regulating valve, 42
Process:
explanation, 33
sequencing, 31, 63
trimming and control, 116 ff.
Process timers, 7, 61
Processing temperatures, 77
Production rates, 196
Projected area, 7
Proportioning temperature controller, 62
Pseudo-plasticity, 33, 84
Purging, 63, 82, 150, 152, 154
Ram, 7
Reciprocating screw machine, 28–31
Relay, 7, 44, 119
Register ring, 7
Re-ground material, 52, 90, 165
Reject mouldings (*see also* Scrap), 165, 191
Re-work, 152, 181
Reynolds number, 94
Rheological properties of plastics, 83, 85
Runner, 7
dimensions and weights, 102
systems, 100, 101
Runners, pressure drop in, 114, 115
Runners, for thermosetting moulding, 127

Safety Council, 13
Safety, 2
devices, 49, 52, 59
spectacles, 155
and housekeeping, 153
Scrap mouldings and profitability, 191, 200
Screw-back time, 31, 134, 194
Screw, 7, 47
special, 48
Screw machine, 29 ff.
Screw motor, 50
Scuffing, 11, 136
Setting (*see also* Trimming the process), 7, 116 ff.
Setting-up, 7, 78
Shear heating, 132, 141
Shear rate:
calculations, 99
derivation and effects on viscosity, 84, 85
in mould filling, 113
Shot capacity, 7
Short moulding, 9, 11
cause and cure, 136
in thermosetting component, 129
Shot weight variation, 117, 146, 147
Shrinkage, 7, 74
Shut-off nozzle, 8

Silvering, silver streak etc., 11
(*see also* Splash mark)
Sink marks, 11, 73, 82, 138
Solenoid valve, 43
Specification criteria, 13
Specific volume of plastics, 16, 72
Spiral flow moulding, 82
Splash mark, 11, 139, 140
Splay mark, 11, 141
Sprue, 8, 90
and runner, for thermosetting mouldings, 127
Sprue bush, 8, 90
Staining, 11
description, cause and cure, 142
of mould, with thermosetting materials, 129
Starting up a machine, 63
Sticking, 11, 142
Stock record, 160
Stores, 180
Stress-free mouldings, 132
Stress, residual or frozen-in, 67, 82, 116, 143
Stripper plate, 8, 95
Stuffing machine, 19
Suck-back, 31, 131
Surface finish, 15
Swelling ratio (after shear), 86
Swept volume, 8
Syndiotactic, 68
Temperature controllers, 62
Temperatures:
for moulding thermosetting materials, 125, 126
processing and decomposition of plastics, 77
Terpolymer, 8
ABS, 70
Thermal diffusivity, 77
Thermal expansion, 80
Thermocouple, 8
placing of in cylinder, 46, 47, 62
Timing devices, 61
Toggle locking, 53–55
Tolerance, 8, 13
Torpedo, 8
Two-cylinder machine, 25

Vented barrel, 8, 48, 49
Ventilation, 154
Vents and venting, 8
Venting of moulds:
for thermoplastics, 95, 131
for thermosetting materials, 128
Vertical flash mould, 91
Viscosity, 33
of molten plastics, 84, 85
effect of shear rate on, 86
Voids, 10, 11, 73
Volume changes on heating of plastics, 73

Warping, 10, 11, 143
Weld lines, 11, 144
Witness marks, 11, 146